PROTOCOLS IN ROOT-MICROBE INTERACTIONS

PROTOCOLS IN ROOT-MICROBE INTERACTIONS

ANJALI S. IYER-PASCUZZI

PURDUE UNIVERSITY PRESS

WEST LAFAYETTE, INDIANA

Cataloging-in-Publication Data on file at the Library Congress.

978-1-62671-281-2 (paperback)
978-1-62671-282-9 (epdf)

Cover image: A group of screenshots taken in a microspore. Courtesy of the author.

CONTENTS

PREFACE AND ACKNOWLEDGMENTS

This book provides protocols for laboratory analysis of root-microbe interactions at the cell, tissue, and organ scales. These protocols were either developed or optimized in the Iyer-Pascuzzi lab for root systems. While these protocols focus on tomato roots, they can be adapted for roots of any plant. When I have included protocols that were optimized from other laboratories, I have included the literature reference for the original protocol.

As with any biology experiment, many people over years contributed to these protocols, and the final protocol in this book builds on the work of many students and postdocs in the Iyer-Pascuzzi lab. The listed contributor is the student or postdoc who most recently successfully used a given protocol. I am grateful to all of the lab members who have worked in the lab over the last 12 years and to the Department of Botany and Plant Pathology at Purdue.

INTRODUCTION

OVERVIEW OF THE PLANT IMMUNE SYSTEM

Plants, unlike animals, lack mobile immune cells and an adaptive immune system. Instead, every plant cell relies on a sophisticated multilayered innate immune system to detect and respond to pathogens such as bacteria, fungi, viruses, and insects (Khan et al., 2025; Jones et al., 2024). This system is essential for plant survival and global food security, as it allows plants to resist a wide variety of biotic threats (Khan et al., 2025; Ali et al., 2024; Jones et al., 2024; Yu et al., 2024; Li et al., 2020).

The first line of defense involves cell-surface receptors known as pattern recognition receptors, which include receptor-like kinases and receptor-like proteins. Pattern recognition receptors detect conserved microbial or pathogen-associated molecular patterns and damage-associated molecular patterns in the extracellular space of plant cells (Khan et al., 2025; Ali et al., 2024; Jones et al., 2024; Ngou et al., 2022). Upon ligand recognition, these receptors trigger signaling cascades, leading to rapid defense responses such as the production of reactive oxygen species, antimicrobial compounds, and cell wall reinforcement (Khan et al., 2025; Ali et al., 2024; Ngou et al., 2022). This branch is effective against a broad spectrum of pathogens, as the recognized patterns are essential for pathogen survival and thus evolve slowly.

The second branch is mediated by intracellular immune receptors, mainly nucleotide-binding leucine-rich repeat receptors (NLRs) (Khan et al., 2025; Ali et al., 2024; Ngou et al., 2022). NLRs detect pathogen-derived effectors either directly or indirectly by sensing modifications to host proteins (Khan et al., 2025; Ali et al., 2024; Benthem et al., 2020). Activation of NLRs results in a robust localized response, including programmed cell death (the hypersensitive response) and downstream defense signaling to contain the pathogen (Ngou et al., 2022). Defense signaling involves complex networks of hormones (e.g., salicylic acid, jasmonic acid, ethylene), transcriptional reprogramming, and systemic signals that can prepare distant tissues for potential attack (Ali et al., 2024; Khan et al., 2025; Nabi et al., 2024; Yu et al., 2024, Bentham et al., 2020). The branches of immunity are interconnected and can amplify each other's signals, providing a flexible and robust defense network (Ali et al., 2024; Khan et al., 2025; Nabi et al., 2024; Yu et al., 2024). The immune system is dynamic and integrates developmental cues and environmental factors to modulate responses throughout the plant life cycle (Li et al., 2020; Ngou et al., 2022).

Understanding plant immunity has major implications for agriculture, as breeding or engineering crops with enhanced immune receptors can improve disease resistance and reduce reliance on chemical pesticides (Jones et al., 2024; Wang et al., 2025). New molecular and cellular knowledge in the plant-microbe field has enabled the transfer and stacking of immune receptors across species, which offers new opportunities for sustainable crop protection (Jones et al., 2024; Wang et al., 2025). This knowledge was enabled by molecular biology and biochemical, cellular, and developmental biology assays such as those described and cited in this book (Wang et al., 2025).

WORKS CITED

Ali, S., Tyagi, A., & Mir, Z. A. (2024). Plant immunity: At the crossroads of pathogen perception and defense response. *Plants, 13,* 1434. https//doi:10.3390/plants13111434.

Bentham, A. R., De la Concepcion, J. C., Mukhi, N., Zdrzałek, R., Draeger, M., Gorenkin, D., Hughes, R. K., & Banfield, M. J. (2020). A molecular roadmap to the plant immune system. *Journal of Biological Chemistry, 295,* 14916–14935. https//doi:10.1074/jbc.REV120.010852.

Jones, J. D. G., Staskawicz, B. J., & Dangl, J. L. (2024). The plant immune system: From discovery to deployment. *Cell, 187,* 2095–2116. https//doi:10.1016/j.cell.2024.03.045.

Khan, M., Srivastava, A. K., Nizamani, M. M., Asif, M., Kamran, A., Luo, L., Yang, S., Chen, S., Li, Z., & Xie, X. (2025). The battle within: Discovering new insights into phytopathogen interactions and effector dynamics. *Microbiological Research, 298*, 128220. https//doi:10.1016/j.micres.2025.128220.

Li, P., Lu, Y. J., Chen, H., & Day, B. (2020). The lifecycle of the plant immune system. *CRC Critical Reviews in Plant Science, 39*, 72–100. https//doi:10.1080/07352689.2020.1757829.

Nabi, Z., Manzoor, S., Nabi, S. U., Wani, T. A., Gulzar, H., Farooq, M., Arya, V. M., Baloch, F. S., Vlădulescu, C., Popescu, S. M., & Mansoor, S. (2024). Pattern-triggered immunity and effector-triggered immunity: Crosstalk and cooperation of PRR and NLR-mediated plant defense pathways during host-pathogen interactions. *Physiology and Molecular Biology of Plants, 30*, 587–604. https//doi:10.1007/s12298-024-01452-7.

Ngou, B. P. M., Ding, P., & Jones, J. D. G. (2022). Thirty years of resistance: Zig-zag through the plant immune system. *Plant Cell, 34*, 1447–1478. https//doi:10.1093/plcell /koac041.

Wang, D., Yang, R., Liu, M., Li, H., Yuan, W., & Zhang, H. (2025). Recent advances in innovative strategies for plant disease resistance breeding. *Frontiers in Plant Science, 16*, 1586375.

Yu, X. Q., Niu, H. Q., Liu, C., Wang, H. L., Yin, W., & Xia, X. (2024). PTI-ETI synergistic signal mechanisms in plant immunity. *Plant Biotechnology Journal, 22*, 2113–2128. https//doi:10.1111/pbi.14332.

PART 1

PROTOCOLS FOR PATTERN-TRIGGERED IMMUNITY (PTI) ASSAYS IN ROOTS

Each protocol in this part has been optimized for tomato (any genotype), but each can be used for roots of other species with small modifications.

1.1

TOMATO ROOT–CHEMILUMINESCENT ASSAY FOR DETECTION OF REACTIVE OXYGEN SPECIES

CONTRIBUTOR: REBECCA LEUSCHEN-KOHL

TIMELINE

Day -2: Media preparation

Day -1: Seed sterilization, plating, and covering

Day 0: Uncover plates and place in light

Days 1–4: Growth

Day 5: Cut, weigh, and recovery

Day 6: Chemiluminescent reactive oxygen species (ROS) assay

PROTOCOL

DAY -2: MEDIA PREPARATION

1% water agar:

Note: This media may be made in large quantities and stored in the 4°C refrigerator if sealed, preventing contamination. 500 mL = roughly 10–15 plates.

MATERIALS

10 g of agar
1,000 mL of ddH$_2$O
150 mm X 150 mm X 10 mm square plates

PROCEDURE

1. In a 2L Erlenmeyer flask, place 10 g of agar and 1,000 mL of ddH$_2$O.
2. Pour plates in plant hood. Let cool/set.
3. Place in a ziplock bag and place in 4°C fridge.

DAY -1: SEED STERILIZATION, PLATING, AND COVERING

SEED STERILIZATION MATERIALS

1.5 mL Eppendorf tube
50% bleach solution
Tomato seeds in 15 mL tubes
Sterile water

SEED STERILIZATION PROCEDURE

1. Add 50% bleach until it covers all seeds, and invert the tube 5 times to mix well.
2. Place the tube on the mixing table for 10 minutes.
3. In the hood, remove the bleach with a pipette and add sterile water.
4. Invert the tube ~5 times and remove the water. Repeat this step until the bleach smell is gone.
5. Fill the tube with sterile water and place it in the 4°C fridge overnight. (Note: Seed can be stored for a few days in the refrigerator; however after 2 days, the timing of germination may be slowed.) <u>Alternative: The seeds can also be plated before placing in the 4°C overnight.</u>

PLATING MATERIALS

Water agar plates
Ruler
Sharpie
Sterilized seeds

PLATING PROCEDURE

1. Unwrap plates in the plant hood. Take the Sharpie and align a ruler at the top horizontally. Draw a line across the plate (roughly 20 cm from the top).
2. Using forceps, place 10 seeds with the apex on the horizontal line on the corresponding side of the plate. We usually plate 4 seeds/genotype and use multiple plates (number depends on the experiment).
3. Seal plates with micropore tape. Wrap plates in foil and place in the 4°C fridge overnight.

DAY 0: UNCOVER PLATES AND PLACE IN LIGHT

MATERIALS

Covered water agar plates with seeds

PROCEDURE

1.24 hours after wrapping the water agar plates with foil, uncover the plates and place them on the racks in the tissue culture room with the face (no-Sharpie side) toward you.

DAYS 1–4: ROOT GROWTH

1. Each day, rotate the plates around the rack to decrease light and temperature effects on each plate.

DAY 5: CUTTING, WEIGHING, AND OVERNIGHT RECOVERY

1. In a 96-well plate, distribute 200 μL of ddH_2O into each well using the multichannel pipette.
2. On the plates of tomato roots, label 30 roots of similar shape/size. We want the roots to be in a similar developmental phase.
3. Using a miniscalpel, cut the tomato plant at the root/shoot junction.
4. Using the sensitive scale, measure the mass of the root. Before measuring, place a small weigh boat on the scale and zero it out, then make a grid or print a template to represent the 96-well plate.
5. Record the mass of the root to the corresponding well in your labeled chart.

6. Carefully place the root into the proper well.

7. Continue to cut, weigh, record, and place into the well for all necessary roots.

8. Once finished, wash the roots using the multichannel pipette to remove the 200 µL of ddH$_2$O and replace with 200 µL fresh ddH$_2$O. Repeat 3X.

9. Cover the plate with aluminum foil and leave in a drawer overnight.

DAY 6: CHEMILUMINESCENT ASSAY FOR DETECTION OF ROS SPECIES

MATERIALS: PREPARATION OF ELICITOR MIXES

Do not add LSS or HPSS until ready to do the assay.

1. MOCK ELICITOR MIX (12 SAMPLES, PREPARE 14X)

	1X (µL)	15X (µL)	NOTES
ddH$_2$O	198	2,970	
Total	198	2,970	
500X LSS	1	15	
500X HPSS	1	15	
Total	200/well	3,000/treatment	

2. CHITIN ELICITOR MIX

	1X (µL)	15X (µL)	NOTES
1 mM chitin stock	0.2	3	Final [chitin] = 1 µM
ddH$_2$O	197.8	2,967	
Total	198	2,970	
500X LSS	1	15	
500X HPSS	1	15	
Total	200/well	3,000/treatment	

3. FLG22 ELICITOR MIX

	1X (µL)	15X (µL)	NOTES
1 mM flg22 stock	0.2	3	Final [flg22] = 1 µM
ddH$_2$O	197.8	2,967	
Total	198	2,970	
500X LSS	1	15	
500X HPSS	1	15	
Total	200/well	3,000/treatment	

PROGRAM SETUP FOR PLATE ASSAY (SET UP ~20 MIN BEFORE STARTING)

- Machine used: Tecan Infinite 200 Pro
- Plate definition: PE 96-well fw—Opti plate—Perkin Elmer—96 White
- Select wells used, highlighted in yellow
- Kinetic cycle: 31 cycles
- Kinetic interval: 2:00
- Luminescence attenuation: Automatic
- Integration time: 1,500 ms

PROCEDURE: ROS ASSAY

1. Before beginning the plate reader assay, turn on the plate reader for 20 min.
2. While the plate reader is warming up, prepare the elicitor mixes (see below), without including the LSS or HPSS mix.
3. Bring gloves, the multichannel pipette, p200 tips, p20 pipette, the waste container, square trays (1/treatment), and HPSS/LSS solutions (on ice) to the reader room.
4. Using the multichannel pipette, remove water from the 96-well plates.
5. Do the following as fast as possible:
6. Add LSS and HPSS to each premade solution, swirl to mix, pour into the square plate, and distribute into the corresponding wells.
7. Repeat for each treatment until all treatments are finished. *Try to only take 1 min total.*
8. Press start on the ROS plate reader program. The program will give you an output of relative light unit (RLU) values at each timepoint (Excel spreadsheet).

DATA ANALYSIS

The data from the Tecan Infinite Pro 200 will come out as an Excel sheet with RLU values for each well at each given time point (Excel spreadsheet).

PROCESS THE DATA

1. For each well, standardize the RLUs by dividing each value by the corresponding root mass (mg).

PLOT THE DATA

Option 1: RLU/mg tissue over time

1. Using processed data, plot the average RLU/mg root tissue for each treatment at each timepoint. See Leuschen-Kohl et al. (2024) for an example.

Option 2: Total RLU/mg tissue (area under the curve)

1. Using processed data, calculate the sum of all RLU/mg tissue values in each time frame for a single well, then multiply by how many minutes the readings are apart. Example: Take the sum of RLU/mg tissue values for cell 1D for 0–60-minute time points, then multiply by 2 because your reader measured every two minutes. See Leuschen-Kohl et al. (2025).

1.2

CHEMILUMINESCENT ASSAY FOR DETECTION OF PHOSPHORYLATED MITOGEN-ACTIVATED PROTEIN KINASE IN ROOTS

CONTRIBUTOR: REBECCA LEUSCHEN-KOHL

TIMELINE

Day -2: Media preparation

Day -1: Seed sterilization, plating, and covering

Day 0: Uncovering

Days 1–4: Growth

Day 5: Cut, recovery, treatment

Day 6/7: Protein extraction, gel, **mitogen-activated protein kinase** (MAPK) antibody, stripping by low pH

Day 7/8: Actin antibody

DAY -2 THROUGH DAY 5: FOLLOW THE PROTOCOL AS IN CHAPTER 1.1

DAY 5: ROOT CUTTING, RECOVERY, TREATMENT, COLLECTION

CUTTING AND RECOVERY

1. In a 6-well plate, distribute 500 μL of ddH$_2$O into each well using a pipette.
2. Using a miniscalpel, cut the tomato plant at the root/shoot junction.
3. Carefully place the root into the proper well. Six roots of the same genotype should be in each well, three wells for each genotype (Mock, flg22, flgII-28).
4. Continue to cut and place into the well(s) for all necessary roots.
5. Once finished, wash the roots by removing the 500 μL of ddH$_2$O and replace with 500 μL fresh ddH$_2$O. Repeat every hour for 6 hours. At 6 hours, the roots can be treated and collected as below.

TREATMENT AND COLLECTION

MATERIALS

Roots, *post 6-hour recovery*
2.0 mL Eppendorf tubes
Sharpie
Liquid nitrogen and small dewar
Forceps (+ethanol)
Pipette + tips
Timer
Elicitor mixes (*see below*)

PREPARATION OF ELICITOR MIXES

Note: Avoid trying to make an elicitor mix that requires less than 1 μL of the pathogen-associated molecular pattern (PAMP) stock. Size up if needed.

MOCK ELICITOR MIX (1X PER GENOTYPE, PREPARE (1X+1))

	1X (μL)	2X (μL)	NOTES
ddH$_2$O	500 μL	1 mL	
Total	500/well	1 mL	

FLG22 ELICITOR MIX

	1X (μL)	2X (μL)	NOTES
1 mM flg22 stock	0.5 μL	1 μL	Final [flg22] = 1 μM
ddH$_2$O	499.5 μL	999 μL	
Total	500	1,000 uL	

FLGII-28 ELICITOR MIX

	1X (μL)	2X (μL)	NOTES
0.5 mM flg22 stock	1 μL	2 μL	Final [flgII-28] = 1 μM
ddH$_2$O	499 μL	998 μL	
Total	500	1,000 μL	

PROCEDURE

1. Prepare Eppendorf tubes labeled with genotype and PAMP for each genotype*treatment combination.

Note: Do not do all treatments and collections at the same time. Spread them out, typically I start two wells every 5 minutes.

1. When ready, remove the 500 μL of water from the desired well and replace with 500 μL of chosen elicitor mix, immediately starting a timer for 10 minutes.
2. At 10 minutes, quickly use ethanol-sterilized forceps to pick up root samples, place them in the corresponding Eppendorf tube, and place them in the liquid nitrogen. Do this as fast as possible.
3. Tissue can be immediately processed or can be stored at -80°C for later analysis.

DAY 6/7: WESTERN BLOT FOR THE DETECTION OF MAPK PHOSPHORYLATION

There are four major steps:

1. Protein extraction and Bradford assay to determine concentration
2. SDS-PAGE
3. Transfer to membrane
4. Antibody treatment

PROTEIN EXTRACTION

MATERIALS

DPDS solution: 0.022 g/500 µL methanol

Protease inhibitor cocktail

Protein extraction buffer: 50 mM Tris-HCl [pH 7.5], 150 mM NaCl, 0.1% Triton X-100

Small pestles (that fit into Eppendorf tubes)

Liquid nitrogen

Tube rack

1.5 mL tubes

Laemmli buffer with beta-mercepthanol: 450 µL 4X Laemmli buffer with 50 µL beta-mercepthanol

- 9:1 ratio. You'll need 25 µL per sample.

Sharpies, preferably two colors (I use red and blue)

PROCEDURE

1. Prepare protein extraction solution: Protein extraction buffer with 1% DPDS solution and 1% protease inhibitor cocktail. Store on ice.
2. Set centrifuge to 4°C; set heating block to 95°C.
3. Prepare TWO sets of Eppendorf tubes, one labeled in red and one labeled in blue, matching the treatments you are processing.
4. One at a time, remove tubes from liquid nitrogen and place in the tube rack. Allow to thaw for 5–10 seconds while simultaneously dipping your tiny mortar into liquid nitrogen.

5. Place the cold mortar into Eppendorf with root tissue and grind until a fine powder. <u>Grind no more than 30 seconds.</u>

6. Immediately add 200 µL of protein extraction solution to the tube and use the mortar to mix. Place on ice.

   a. Repeat Steps 4 and 5 for each sample.

7. Once all samples have been ground, centrifuge at 10,000 g for 10 minutes at 4°C.

   a. Carefully remove from the centrifuge and place back on ice, being sure not to agitate the pellet.

   b. While this is centrifuging, I usually set up for the **Bradford assay** and make the **Laemmli buffer.**

8. Transfer 100 µL of the supernatant to the prelabeled red 1.5 mL Eppendorf tubes.

   a. From the 100 µL, transfer 75 µL of solution to the blue 1.5 mL Eppendorf tubes.

   b. Move forward with protocol using blue tubes. The red tubes will be used for the Bradford assay (see next section).

9. Add 25 µL of Laemmli solution (with BME) to samples in blue tubes and mix briefly.

   a. Note: This is where you now must think about your protein concentration from the Bradford assay being 75% of your protein sample. Example: Your Bradford assay shows a concentration of 150 ng/µL. You have 112.5 ng/µL.

$$[New] = [Original] \times \frac{Original\ Volume}{Total\ Volume}$$

10. Boil blue tubes for 10 minutes at 95°C. Samples are now ready for loading into the **SDS-PAGE gel.**

BRADFORD ASSAY TO DETERMINE PROTEIN CONCENTRATION BEFORE SDS-PAGE

MATERIALS

Bradford solution

Bovine serum albumin

Protein samples from extraction, red tubes

Spectrophotometer (A595)

Cuvettes

Pipette + tips

PROCEDURE

1. Prepare a Bradford standard curve using bovine serum albumin per manufacturer's instructions.

    a. We added 10 µL of each concentration to 1,500 µL of Bradford solution, so we will do the same for the determination of protein concentration.

2. Prepare cuvettes with 1,500 µL of the Bradford solution, one for each sample and one for the blank.

3. To each corresponding cuvette, add 10 µL of the protein solution from the red tubes. Cover with parafilm, invert to mix, and let rest 15 minutes at room temp.

4. Measure the absorbance of all samples at 595 nm.

5. Using absorbance values and standard curve, determine protein concentration of each sample.

    a. <u>Don't forget that this is diluted upon the addition of Laemmli buffer in the blue samples.</u>

SDS-PAGE GEL

MATERIALS

10% SDS-PAGE polyacrylamide gel

Pipette

Fine-point pipette tips for loading

10X Tris/Glycine/SDS (1 L final volume): 30 g Tris, 144 g glycine, 10 g SDS in ddH$_2$O

1X Tris/Glycine/SDS (1L final volume): 100 µL 10X Tris/Glycine/SDS, 900 mL ddH$_2$O

PROCEDURE

1. Prepare the SDS-PAGE apparatus with 1X Tris/Glycine/SDS and 10% acrylamide gel.

2. Using concentration from the Bradford assay, determine the volume needed for 2,000 µg of protein.

3. From left to right, add 5 µL of protein ladder to a well and then to each protein sample, being sure to load each well with the correct volume of protein.

4. Run gel at 130 V for 10 minutes and then at 80 V for 1.5 hours until the ladder has started to run off of the bottom of the gel.

5. Continue to protocol for **transfer to membrane.**

TRANSFER TO MEMBRANE

MATERIALS

SDS-PAGE gel, after samples have been run

Nitrocellulose membrane

Shallow container

Transfer buffer solution (TBS) (1 L final volume): 100 mL of 10X Tris/Glycine buffer (premade) + 200 mL methanol + 700 mL dH_2O

PROCEDURE

1. Carefully separate the glass plates of the acrylamide gel and use a razor to cut out the area of interest for transfer. <u>Be sure not to let the gel dry; keep it in a shallow container of TBS.</u>

2. Cut a section of nitrocellulose membrane large enough to cover the entire membrane.

3. Assemble the transfer sandwich, making sure there are NO AIR BUBBLES. Keep everything wet; do not let it dry out.
   a. Black side (anode), sponge, blotting paper, gel, membrane, blotting paper, sponge, red side (cathode).

4. Place the sandwich in the chamber, add the ice pack, and fill with TBS to indicated line.

5. Transfer total proteins to the membrane at 100 V for 50 minutes.

PONCEAUS STAIN (TO CHECK IF PROTEINS TRANSFERRED CORRECTLY)

MATERIALS

PonceauS Stain

Square petri dish or similar

Membrane, after completed transfer

10X TBS

1X TBST: 100 mL of 10X TBS + 899 mL ddH_2O + 1 mL Tween20

PROCEDURE

1. Place the membrane in a petri dish and add enough stain to cover. Place on rocker for 10 minutes.
2. After 10 minutes, carefully pick up the membrane with forceps and wash with ddH_2O gently until lanes are visible.
3. Image the gel.
4. Place the gel in a petri dish with TBST and rock for 10 minutes. Remove the TBST and replace with fresh. Repeat for a total of four washes or until stain is fully removed.

BLOCKING, MAPK ANTIBODY, DETECTION

Note: You must perform the MAPK antibody detection first, then Actin. This is because the MAPK gets de-phosphorylated in the stripping step.

MATERIALS

Square petri dish or similar
Membrane
1X TBST
5% skim milk blocking solution: 2.5 g in 50 mL TBST
MAPK antibody: Phospho-p44/42 MAPK (Erk1/2) (Thr202/Tyr204) (D13.14.4E) XP® Rabbit mAb (HRP Conjugate) #8544

PROCEDURE

1. Place the membrane in a petri dish and cover with 10 mL blocking solution for 1 hour at room temp (or overnight at 4°C). Place on rocker.
2. Discard the blocking solution and wash the membrane with 10 mL TBST. Rock for 5 minutes. Discard the TBST and refresh with 10 mL TBST. Repeat for a total of 3 washes.
3. To 10 mL blocking solution, add the antibody per manufacturer's instructions.
4. Place the membrane in a solution containing the antibody and rock for 1 hour at room temperature or overnight at 4°C.
5. Discard the solution and wash the membrane with 10 mL TBST and rock for 5 minutes. Discard the TBST and refresh with 10 mL TBST. Repeat for a total of 3 washes.

6. Prepare a 1:1 (v/v) solution of detection substrate. <u>Turn on the machine before adding the detection substrate to the membrane.</u>

7. Place the membrane in a clear plastic wrap or a clear ziplock bag. Take to the reader.

8. Using the pipette, put the substrate on the protein side of the membrane and cover with plastic.

9. Transfer to the reader and take an image.

STRIP THE MEMBRANE USING LOW PH
(TO REMOVE ANTIBODY FROM MEMBRANE
IN ORDER TO USE A DIFFERENT ANTIBODY)

MATERIALS

Square petri dish

1X TBST

Stripping solution (250 mL total volume): 3.75 g glycine, 25 mL 20% SDS, 2.5 mL Tween 20 in ddH_2O, pH to 2.2 with HCl

PBS Solution (500 mL total volume): 4 g NaCl, 0.1 g KCl, 0.72 g Na_2HPO_4, 0.12 g KH_2PO_4 in 400 mL ddH_2O. pH to 7.4 with HCl. Fill to final volume of 500 mL

PROCEDURE

1. Place the membrane in a petri dish and add 10–15 mL of stripping solution. Rock for 10–15 minutes but no more than 15!

2. Discard the buffer, add 10 mL PBS solution, and rock for 10 minutes.

3. Discard the solution and wash the membrane with 10 mL TBST. Rock for 5 minutes. Discard the TBST and refresh with 10 mL TBST. Repeat for a total of 2 TBST washes.

4. Move forward with reblocking and new antibody detection.

BLOCKING, ACTIN ANTIBODY, DETECTION

MATERIALS

Square petri dish, or similar

Membrane

1X TBST

1X TBST: 100 mL of 10X TBS + 899 mL ddH$_2$O + 1 mL Tween20

5% skim milk blocking solution

5% skim milk blocking solution: 2.5 g in 50 mL TBST

Actin antibody: HRP conjugated Anti-Plant Actin Mouse Monoclonal Antibody (3T3) (Abbkine, ABL1055)

PROCEDURE

1. Place the membrane in a petri dish and cover with 10 mL blocking solution for 1 hour at room temp (or overnight at 4°C). Place on rocker.

2. Discard the blocking solution and wash the membrane with 10 mL TBST. Rock for 5 minutes. Discard the TBST and refresh with 10 mL TBST. Repeat for a total of 3 washes.

3. To the 10 mL blocking solution, add the antibody per manufacturer's instructions.

4. Place the membrane in a solution containing the antibody and rock for 1 hour at room temperature or overnight at 4°C.

5. Discard the solution and wash the membrane with 10 mL TBST. Rock for 5 minutes. Discard the TBST and refresh with 10 mL TBST. Repeat for a total of 3 washes.

6. Prepare a 1:1 (v/v) solution of detection substrate.
   a. Turn on the machine <u>before</u> adding the detection substrate to the membrane.

7. Place the membrane in a clear plastic wrap or clear ziplock bag. Take to the reader.

8. Using the pipette, put the substrate on the protein side of the membrane and cover with plastic.

9. Transfer to the reader and take an image.

DATA ANALYSIS

MAPK activation was quantified using an established ImageJ plugin (Ohgane & Yoshioka, 2019). See Leuschen-Kohl et al. (2025) for example images.

1.3

TEMPORARY ROOT GROWTH ASSAY AFTER PAMP TREATMENT

CONTRIBUTOR: REBECCA LEUSCHEN-KOHL

The following is an outline of the experimental setup for whole root phenotypic response to PAMP treatments (including HK bacteria). For best results, optimize the concentration of PAMP to use before doing a large experiment.

Day -2: Media preparation
Day -1: Seed sterilization, plating, and covering
Day 0: Uncovering
Day 1–3: Growth
Day 4: PAMP inoculation
Day 5: Root imaging (24 hpi)
Day 6: Root imaging (48 hpi)

Follow the protocol in Chapter 1.1 for Day -2 media preparation and for Day -1 seed sterilization and plating procedures.

DAY 0: UNCOVER PLATES AND PLACE IN LIGHT

MATERIALS

Covered water agar plates with seeds

PROCEDURE

1. Twenty-four hours after wrapping the water agar plates with foil, uncover the plates and place them on the racks in the tissue culture room with the face (no-Sharpie side) toward you.

DAY 1–3: ROOT GROWTH

PROCEDURE

Each day, rotate the plates around the rack to decrease light and temperature effects on each plate.

DAY 4: INOCULATION DAY

1. On Day 4, each side of the plate was randomly assigned to either mock treatment or PAMP treatment, making sure that both treatments were represented on each plate. Be sure to number the plates for image analysis comparison.

PREPARATION OF PAMP/HK BACTERIA TREATMENT

Note: The same sterile water used to create the PAMP solution should be used for the mock treatment.

1 µM flg22 (10 mL) (or csp22):
10 µL of 1 mM flg22 stock solution
9.90 mL of sterile water
1 µM flgII-28 (10 mL):
1 µL of 1 mM flgII-28 stock solution
9.90 mL of sterile water
100 nM flgII-28 (10 mL):
10 µL of 1 mM flgII-28 stock solution into 90 mL sterile water (100 µM)
10 µL of 100 µM flgII-28 stock solution into 9.90 mL of sterile water

TREATMENT OF TOMATO ROOTS WITH PAMP SOLUTION/ CONTROL: GROUP 1

MATERIALS

Sterile water

Elicitor: 1 µM flg22 Treatment Solution

    1 µM or 1oo nM flgII-28 Treatment Solution

Plates with tomatoes, labeled

PROCEDURE

1. **Before inoculating**, scan the plates using the WhinRhizo imager. **Make sure** to have a ruler present in each of the scans for scale, and be sure to place a dot next to where the root tip is at the 0 hpi scan. *This dot will not be used for quantitative analysis, just for reference.*

2. In the hood, remove the tape from the plates.

3. For those plates marked with PAMP treatment, use a pipette to dispense 300 µL of PAMP treatment or sterile water over root organs on the respective halves of the plate.

4. Let the plates air-dry in the hood momentarily (until the liquid is mostly evaporated) before rewrapping and placing them back onto the light racks.

DAYS 5–7: ROOT IMAGING

1. **At 24, 48, and 72 hours postinoculation**, scan the plates using the WhinRhizo imager. **Make sure** to have a ruler present in each of the scans for scale, and be sure to place a dot next to where the root tip is at the 24, 48, or 72 hpi scan. Again, *this dot will not be used for quantitative analysis, just for reference.*

2. **Analysis**: In ImageJ, use the ruler in your scans to set your scale. Using the measurement tool, measure the length of the root from the 0 hpi scan, the 24 hpi scan, 48 hpi scan, etc. Make sure to record in an Excel file.

Notes: Be sure to label plates well. As an example:

RGI_1uMflg22_mock_H7996_0hpi_Plate1_12.08.2022.tiff

This can be read RGI assay, with treatments 1 μM flg22 and mock, for tomato cultivar H7996. This is Plate 1 and was scanned at 0 hpi on 12.08.2022.

1.4

GENERATING HEAT-KILLED BACTERIA

CONTRIBUTOR: REBECCA LEUSCHEN-KOHL

The following is an outline of the experimental setup for generating heat-killed (HK) *Ralstonia*. This can be modified for other microbes. HK microbes are useful for determining the root phenotypic response to microbial PAMPs in a root ROS assay (see Chapter 1.1), MAPK assay (see Chapter 1.2), or root growth inhibition assay (see Chapter 1.3).

Day -2: Media preparation
Day 0: Preparation of HK *Ralstonia*
Day 2: Measurement of CFU/mL

DAY -2: MEDIA PREPARATION

Note: This media may be made in large quantities and stored in the 4°C refrigerator if sealed to prevent contamination.

PREPARE CPG PLATES (FINAL VOLUME: 500 ML)

MATERIALS

2.5 g Glucose
0.5 g casamino acid
0.5 g yeast extract

5 g peptone
8 g agar (placed in a flask that will be used to autoclave)
1 mL of 1 % TZC (once autoclaved and cooled)

PROCEDURE

1. Using a large glass beaker and a stir bar, add the weighed materials to 450 mL ddH$_2$O.
2. Turn on the stir plate and allow the media to mix for 5–10 minutes.
3. Use the pH meter, adjusting the pH to be between 6.5 and 7.
4. Pour the solution into a graduated cylinder and add sterile water until the final volume is 500 mL. Briefly place the water back into the plastic beaker and stir.
5. Pour the solution from the beaker into the bottle with the agar.
6. Place the bottle with media solution into the autoclave.
7. After autoclaving, allow the media to cool and then add 1 mL of 1% TZC. Note: While cooling, turn on the hood and sterilize.
8. Pour the plates and allow them to cool before placing in a 4°C fridge or use for plating.

PLATING *RALSTONIA* (~48 HOURS BEFORE THE TIME OF INOCULATION)

MATERIALS

CPG plates with 1% TZC (and selection antibiotics if necessary)
Ralstonia glycerol stock from -80°C, stored in dewar with liquid nitrogen
Glass spreader
Ethanol
Flame/fire source
Pipette + tips

PROCEDURE

1. In the bacterial hood, flame-sterilize the glass spreader and allow it to come to room temperature.
2. Using the pipette, deposit 200 μL of *Ralstonia* from a glycerol stock on the CPG. Spread *Ralstonia* lawn using the glass spreader.

3. Flame-sterilize the spreader to kill *Ralstonia*.

4. Place labeled *Ralstonia* + CPG plates in 28°C for 48 hours.

DAY 0: PREPARATION OF HEAT-KILLED *RALSTONIA*

Note: This procedure can be done for use in ROS, root growth inhibition assay, etc. The root growth inhibition assay is written here.

MATERIALS

Ralstonia plates
ddH$_2$O
CPG plates
Sharpie
PCR tubes or plate
15 mL Falcon tubes
Pipette + tips
50% bleach solution (for waste)
Sterile glass slide
70% ethanol
    Heat block at 95°C

PROCEDURE—RESUSPENSION

1. Sanitize the hood with 70% ethanol, and place water and empty 1L bottles into hood.

2. Make a solution of 30–50% bleach to place contaminated instruments (glass slides, pipette tips, etc.) in.

3. Set up Eppendorf tubes for titer (CFU/mL determination), typically 8x serial dilutions (10^{-1} through 10^{-8}),

4. Remove inoculated petri dishes from the incubator and place into the hood.

5. Gently scrape *Ralstonia* with a sterilized glass slide and place it in a 15 mL Falcon tube with 10 mL ddH$_2$O. Cap and gently shake to resuspend. This is the initial inoculum.

6. Place the glass slide into the bleach solution.

7. Take the inoculum to the spectrophotometer to get optical density (OD). Adjust until desired OD is reached. (We use an OD that corresponds to 10^8 CFU/mL; this can vary by spectrophotometer and should be empirically determined.)

8. Once the target OD is reached, place 1.5 mL into an Eppendorf tube for the titer protocol.

PROCEDURE—TITER PROTOCOL

1. Set up PCR tubes or a 96-well plate and label from 10^{-1} to 10^{-8}.

2. Fill each well with 180 μL of water.

3. Starting with the inoculum from the resuspension protocol (OD=0.2), transfer 20 μL to the tube labeled -1. Mix. Transfer 20 microliters from the tube labeled -2. Mix. Repeat steps until 20 microliters of the tube labeled -7 has been transferred to the tube labeled -8.

4. Plate 20 microliters of the -4, -5, -6, and -7 dilutions onto CPG plates.

5. After 2 or 3 days, count the number of colonies and calculate CFU/mL.

PROCEDURE—HEAT-KILL *RALSTONIA*

Note: At this point, you will be killing the bacteria. **Be sure to do the titer BEFORE killing the bacteria.** Then, plate the bacteria after heat to ensure death.

1. After reserving inoculum for bacterial titer, place enough inoculum into 1.5 mL tubes for your experiment plus additional for plating to ensure bacterial death.

2. Place 1.5 mL tubes in heat block at 95°C for 10 minutes.

3. After 10 minutes, remove tubes and pool samples. Take 100 μL of the heat-killed solution and plate on CPG. Place in 28°C and check for any colonies 48 hours later.

4. The rest of the inoculum is now ready to use. You can use the bacteria for any assay. The ROS assay is described in Chapter 1.1., MAPK is described in Chapter 1.2, and root growth-inhibition assay is described in Chapter 1.4.

DAY 2: MEASUREMENT OF CFU/ML

1. Using dilution plates, determine the CFU/mL of inoculum.
2. Examine CPG plated with heat-killed bacteria. No colonies should appear if heat-kill was successful.

1.5

PAMP TREATMENT FOR RNA-SEQ IN ROOTS

CONTRIBUTOR: REBECCA LEUSCHEN-KOHL

This is the procedure we use for examining gene expression after PAMP/HK bacteria treatment in roots. Note that you can use this for whole roots or to examine different developmental regions of roots.

Day -2: Media preparation
Day -1: Seed sterilization, plating, and covering
Day 0: Uncovering
Days 1–4: Growth
Day 5: Cut, recovery (this is if you are examining root developmental regions)
Day 6: Treatment, collection
Day 7: RNA extraction

Follow earlier chapters for water agar preparation, seed sterilization, plating, and covering.

SEED PLATING

Follow directions in earlier chapters.

Seeds were sterilized, and then enough seeds were plated at 10 seeds/plate/genotype for all necessary treatments. Seal plates with micropore tape. Wrap plates in foil and place in 4°C overnight.

EXPERIMENTAL DESIGN

This will vary by experiment. You will want both mock and treated plants on the same plate and multiple plates. You will have to optimize the number of plates and plants prior to doing the experiment. An example in which we are cutting different developmental regions of the root is below.

PLATING EXAMPLE

TREATMENTS

Mock, flg22, flgII-28

SEGMENTS

Whole root, late differentiation (LD) zone, early differentiation (ED) Zone. *Note that LD and ED samples come from the same root.*

EXAMPLE

6 roots/treatment (~1 plate) * 3 treatments/genotype * 3 reps / genotype * 2 roots (WR, LD/ED) = **18 plates per genotype (+ 2 extra in case of germination issues).** Make sure you use a full balanced experimental design to avoid statistical issues in your RNAseq data analysis.

DAY 0: UNCOVER PLATES AND PLACE IN LIGHT

1. 24 hours after wrapping the water agar plates with foil, uncover the plates and place them on the racks in the tissue culture room with the face (no-Sharpie side) toward you.

DAY 1–4: ROOT GROWTH

1. Each day, rotate the plates around the rack to decrease light and temperature effects on each plate.

DAY 5: ROOT CUTTING AND RECOVERY

MATERIALS

Tomato seedlings
Scalpel
6-well plate
Elicitor mixes
ddH_2O
Pipette + tips

PROCEDURE

1. In a 6-well plate, distribute 500 µL of ddH_2O into each well using a pipette.
2. Using a miniscalpel, cut the tomato plant at the root/shoot junction. For LD/ED samples, place under a light microscope, cut off the meristem (where root hairs are not yet emerging), and cut at the point where root hairs appear fully emerged. See Leuschen-Kohl et al. (2025) for an example.
3. Carefully place the root into the proper well. Six roots of the same genotype should be in each well, three wells for each genotype (Mock, flg22, flgII-28).
4. Continue to cut and place into the well(s) for all necessary roots.
5. Once finished, wash the roots by removing the 500 µL of ddH_2O and replace with 500 µL fresh ddH_2O. Repeat every hour for 3 hours, totaling 3 washes. Cover with foil and leave overnight.

DAY 6: ROOT TREATMENT, COLLECTION

MATERIALS

Roots, *postovernight recovery*
2.0 mL Eppendorf tubes
Sharpie
Liquid nitrogen and small dewar
Forceps (+ethanol)
Pipette + tips

Timer

Elicitor mixes, *see below*

PREPARATION OF ELICITOR MIXES

Note: Avoid trying to make an elicitor mix that requires less than 1 µL of the PAMP stock. Size up if needed.

MOCK ELICITOR MIX

	1X (µL)	3X (µL)	NOTES
ddH$_2$O	500 µL	1,500 µL	
Total	500 µL /well	1,500 µL	

FLG22 ELICITOR MIX

	1X (µL)	3X (µL)	NOTES
1 mM flg22 stock	0.5 µL	1.5 µL	Final [flg22] = 1 µM
ddH$_2$O	499.5 µL	1,498.5 µL	
Total	500 µL	1,500 µL	

FLGII-28 ELICITOR MIX

	1X (µL)	3X (µL)	NOTES
0.5 mM flg22 stock	1 µL	3 µL	Final [flgII-28] = 1 µM
ddH$_2$O	499 µL	1,497 µL	
Total	500 µL	1,500 µL	

PROCEDURE

1. Prepare Eppendorf tubes labeled with genotype and PAMP for each genotype*treatment combination.

   Note: Do not do all treatments and collections at the same time. Spread them out. Typically I start two wells every 5 minutes.

2. When ready, remove the 500 µL of water from the desired well and replace with 500 µL of chosen elicitor mix, immediately starting a timer for 6 hours.

3. At 6 hours, quickly use ethanol-sterilized forceps to pick up root samples, place them in the corresponding Eppendorf tube, and place them in the liquid nitrogen. Do this as fast as possible!

4. Tissue can be immediately processed or can be stored at -80°C for later analysis.

RNA EXTRACTION

RNA extraction and RNA purification was done with Qiagen RNeasy minikit with DNase I on-column treatment, following the instructions in the Qiagen kit.

1.6

QUALITATIVE ANALYSIS OF CALLOSE DEPOSITION IN DEX-INDUCIBLE TRANSGENIC *ARABIDOPSIS THALIANA* ROOTS FOLLOWING PAMP TREATMENT

CONTRIBUTOR: ABIGAIL ROGERS

Note: This protocol has been adapted and optimized from Schenk and Schikora (2015). We find that for roots this assay is best for *qualitative* analyses. You can also use this protocol without the dexamethasone (DEX) in a wild-type plant.

MATERIALS

Plant growth media—0.5x MS + 0.5% Sucrose, pH 5.7 with KOH.

For 1 liter:

2.165 g MS + basal salts

0.25 g MES

5 g sucrose

10 g micropropagation agar (leave out for liquid media)

*Pour ~50 mL solid media into small square plates for growing seedlings.

Sterilization Solution—50% Bleach + 0.1% Tween-20
For 10 mL:
    5 mL Bleach
    5 mL sterile water
    10 µL Tween-20

DEX stock solution—25 mM DEX
For 20 mL:
    0.196 g DEX (Note: Use caution)
    20 mL dimethyl sulfoxide (DMSO)
*Store at -20°C in dark.

Pretreatment solution (plant growth media + 10 µM DEX)
For 1L:
    1L Sterilized liquid plant growth media (see above)
    400 µL 25 mM DEX stock solution
*Prepare fresh for each experiment.

Pretreatment control solution (plant growth media + DMSO)
For 1L:
    1L sterilized liquid plant growth media
    400 µL DMSO
*Prepare fresh for each experiment.

Destaining solution (1:3 Acetic Acid:Ethanol)
For 1L:
    250 mL acetic acid
    720 mL ethanol

Wash solution (150 mM K_2HPO_4, 0.01% aniline blue)
For 50 mL:
    1.3 g K_2HPO_4
    5 mg aniline blue
    50 mL sterile water
*Prepare immediately prior to use.

15 mL Falcon tubes

Laminar flow hood

Small square plates

Micropore tape

6-well tissue culture plates

Aluminum foil

Laser-scanning confocal microscope equipped with a DAPI filter

50% Glycerol for plating samples on microscope slides

Microscope slides and cover slips

PROCEDURE

SEED STERILIZATION AND STRATIFICATION

1. Place *Arabidopsis thaliana* seeds in a sterile Falcon tube with the sterilization solution.

2. Place the tube on a tilt plate and allow the seeds to sterilize for 3–5 minutes.

3. Remove the sterilization solution via pipetting in a laminar flow hood. *If using different *A. thaliana* genotypes, ensure that you do not mix your seeds by replacing the pipette tip after removing any solution from the Falcon tube.

4. Rinse seeds with sterile water 5–6x in the laminar flow hood.

5. Resuspend the seeds in sterile water.

6. Plate seeds on solid plant growth media.

7. Tape plates with micropore tape.

8. Cover plates with aluminum foil and place at 4°C for 48 hours before moving the plates to a light rack. (This stratification step promotes uniform and improved germination.)

9. Following stratification, move the plates to a light rack or growth chamber.

10. Allow seedlings to germinate and grow for seven days.

PROCEDURE (7 DAYS AFTER MOVING PLANTS TO LIGHT)

1. Gently remove the *A. thaliana* seedlings from the plant growth media and move them into pretreatment media (control and DEX treatments).
   a. A six-well tissue culture plate works well for this step.

2. Allow plants to incubate in pretreatment solutions for three days in a growth chamber or on a light rack.
   a. Replace the liquid media daily, as DEX is light-sensitive.

3. Three days after beginning pretreatment, replace the pretreatment media a final time and add PAMP.
    a. For flg22 (100 nM)
4. Incubate plants in pretreatment media with PAMP for 24 hours.
5. After 24 hours of incubation time, replace the PAMP solutions in the tissue culture plates with the destaining solution.
6. Incubate in a chemical fume hood until seedling cotyledons are transparent.
    a. This step typically takes overnight, and you can consider gently shaking the tissue culture plate to speed up this process.
    b. Replace destaining solution 2–3x to improve clearing of the plant tissue.
7. Once cotyledons are destained (appear clear), remove the destaining solution and replace with the wash solution for 30 minutes. Gently agitate the tissue culture plate during this washing step.
    a. Prepare staining solution while seedlings are incubating in the wash solution.
8. Remove the wash solution and add the staining solution to the tissue culture wells.
9. Incubate for 2 hours at room temperature.
    a. Cover the tissue culture plate in aluminum foil as aniline blue is light-sensitive.
10. Seedlings should now be ready to image.

CALLOSE IMAGING USING A LASER-SCANNING CONFOCAL MICROSCOPE

Setting up confocal (we use a Zeiss 880 Upright Confocal Microscope)

1. Select DAPI filter from "smart setup."
    a. Laser setting 0.6, pinhole 1AU. Start with these parameters and adjust as needed. Ensure that all imaging parameters are maintained across experiments.
2. Adjust frame size to 1024 x 1024.
3. Set line averaging to 4 (mode-line).
4. Set lens to 10x-apochromat, air lens.

IMAGING PARAMETERS

1. Adjust the microscope platform so that the root's elongation zone (including the start of the transition to the root tip) is in the center of your viewfinder.
   a. Do this step using brightfield.
2. Adjust the focus so that as many epidermal cells are in focus as possible.
3. Take a snapshot of this focal plane (16-bit depth) and save.
4. At same focal position, select z-project from acquisition parameters.
5. Switch to live view and set the top (uppermost root cells) and bottom (lower root cells) focal planes of the z-stack.
6. Run the z-stack program.
   a. For efficiency purposes, ensure that z-stacks are imaged at the maximum speed at an 8-bit depth. These will create nice z-stacks without taking too much time.
7. Save the z-project.
8. Repeat steps for each seedling.

1.7

CHEMICAL INHIBITION OF THE ACTIN AND MICROTUBULE NETWORKS IN TOMATO SEEDLING ROOTS

CONTRIBUTOR: ABIGAIL ROGERS

This protocol is useful for determining how cytoskeleton disruption impacts developmental and immune pathways.

MATERIALS

LatrunculinB stock (5 mM, store at -20°C)
For 1 mL:
    1 mL DMSO
    1.98 mg LatrunculinB

LatrunculinB working solution (10 µM)
For 5 mL:
    10 µL LatrunculinB stock
    4.99 mL sterile water

Mock-LatrunculinB solution (DMSO)
For 5 mL:

10 μL DMSO
4.99 mL sterile water

Orzyalin Stock (10 mM, store at -20°C)
For 1 mL:
    1 mL DMSO
    3.46 mg Oryzalin

Oryzalin working solution (100 μM)
For 5 mL:
    50 μL Oryzalin stock
    4.95 mL sterile water

Mock-oryzalin solution (DMSO)
For 5 mL:
    50 μL DMSO
    4.95 mL sterile water

1% water/agar
For 1 L:
    10 g agar
    1 L water

50% bleach
Sterile water
50 mL Falcon tube
Medium square plates
Micropore tape
Tomato seeds

PROTOCOL

GROWING TOMATO SEEDLINGS

1. Sterilize tomato seeds in the 50% bleach solution for 15 minutes.
    a. Place on a tilt plate to ensure even sterilization.
2. Rinse the tomato seeds with sterile water 3x in the laminar flow hood.

3. Stratify seeds in sterile water at 4°C overnight.
4. After stratification, plate seeds on 1% water/agar plates.
   a. Plate so that the radicle will emerge toward the bottom of the plate.
5. Secure the plate lid using micropore tape, and move the plates to a light rack.
   a. Place upright so that roots will develop toward the bottom of the plate, along the surface of the media.
6. Allow seedlings to develop for 4–6 days.
   a. Timing is dependent upon how rapidly seedlings develop. This could be influenced by the time of year or the genotype of the tomato.

TREATING WITH CHEMICAL INHIBITORS

1. On treatment day, freshly prepare working solutions of the LatrunculinB and Oryzalin and their respective mock treatments.
2. Remove plates from the light rack and open them in a laminar flow hood.
3. Allow residual water to evaporate.
4. Once the surface of the media appears relatively dry, pipette 200 µL of working solution along the entire length of each root.
5. Allow chemical treatments to dry.
   a. Following this step, you can expose roots to PAMPs (see earlier chapters), pathogens, or other treatments to test how disruption of actin or microtubules influences the root's response to stressors.

1.8

EFFECTOR PROTEIN EXPRESSION *IN PLANTA*

CONTRIBUTOR: ABIGAIL ROGERS

Although the experiment here is a leaf-based protocol, we include this method because it is commonly used in our lab for effector proteins from *Ralstonia solanacearum*. Transient expression of effectors can be used for multiple experiments, including testing the role of effector proteins in suppressing pattern-triggered immunity, for effector protein localization *in planta* or to identify interacting partners of an effector protein as part of a coimmunoprecipitation assay. Here we give the basic protocol for expression and imaging.

To observe effector proteins in roots using the hairy root method, readers are directed to protocols in Ron et al. (2016) and Denne et al. (2021).

MATERIALS

100 mM $MgCl_2$, 500 mL (autoclaved)

    MilliQ water, 500 mL

    $MgCl_2$ (MW: 95.21 g/mol), 4.76 g

    *Adding $MgCl_2$ to water creates an exothermic reaction. The beaker will warm, and vapor may be released. Avoid placing your face directly over the beaker when adding $MgCl_2$ to water.

10 mM $MgCl_2$, 50 mL

    Sterile water, 45 mL

    Autoclaved 100 mM $MgCl_2$ stock, 5 mL

LB (10 mL LB/liquid culture)

    Use standard recipe to make LB

LB + agar (1 plate with selection/construct)

    Use standard recipe to make LBA

100 mM acetosyringone (Filter sterilized)

    1.96 g of acetosyringone (FW = 196.19 g/mol)

    100 mL DMSO

    Combine components and filter-sterilize into 1.5 mL tubes. Store at -20°C and thaw for use.

    ~4-week-old, healthy, *N. benthamiana* plants

PROCEDURE

1. Three days prior to infiltration, streak *Agrobacterium* stocks on LBA + selection plates.

2. Allow to grow at 28–30°C for 2 days.

3. One day prior to infiltration, prepare an overnight culture of *Agrobacteria* by taking a scoop of bacteria from the plate (does not need to be single colony) and inoculating 10 mL of LB + antibiotic selection.

4. Shake overnight at 30°C at ~225 rpm, 16–20 hours (no more than 20 hours, otherwise the bacteria will be in stationary/decay phase).
   - If using a 50 mL Falcon tube, ensure that the cap is not on all the way, as you want to be sure that your culture has sufficient air.

5. On the day of infiltration, spin down 10 mL overnight cultures at ~3.7 rpm for 3 min.

6. Discard supernatant.

7. Resuspend cells in 4 mL of 10 mM $MgCl_2$.

8. Take 1 mL of resuspended cells and use for OD600 reading (make sure to blank the spectrophorometer with $MgCl_2$).

9. A range of 0.5–0.7 for OD600 reading is acceptable. Anything above this will need to be diluted to an OD600 of ~0.5. Anything below this will require spinning down the culture again and resuspending in less $MgCl_2$.
   - Use $C_1{}^*V_1 = C_2{}^*V_2$ to determine dilution of the culture. You do not need to reread the OD after dilution.
   - Will need ~2 mL of culture for infiltration.

10. Once all cultures are ~OD600 of 0.5, add 100 µM of acetosynringone.

- If you have made your stock at 100 mM, add the same number of μL acetosyringone as you have mL of culture (if you have a 5 mL culture, add 5 μL of 100 mM acetosyringone).

11. Allow cultures + acetosyringone to incubate at RT for 2–4 hours (~3 hours).

12. 1–1.5 hours prior to your inoculation time, bring your plants out of their growth chambers and allow them to acclimate to the environment in which you will infiltrate them.

13. Infiltrate *Nicotiana* plants with a 1 mL needleless syringe (we most often use *N. benthamiana*).
    a. Pull up culture to fill the syringe.
    b. Select the leaf to infiltrate (younger and flat leaves are easier to infiltrate).
    c. Place the tip of the syringe on back of the leaf and sandwich the leaf against the syringe by placing your finger on the topside of the leaf, pressing it onto the syringe (firmly enough to create a seal to push your bacteria into the leaf tissue but not so hard that you damage the leaf).
    d. Use the syringe plunger to inject bacteria into leaf tissue.

14. Place plants back in the original growth chamber and use for experiments (confocal microscopy, pattern-triggered immunity assays, co-IP, co-IP/MS, etc).

WORKS CITED

Denne, N. L., Hiles, R. R., Kyrysyuk, O., Iyer-Pascuzzi, A. S., & Mitra, R. M. (2021). *Ralstonia solanacearum* effectors localize to diverse organelles in *Solanum* hosts. *Phytopathology, 111,* 2213–2226. https//doi:10.1094/PHYTO-10-20-0483-R.

Leuschen-Kohl R., Roberts R., Stevens D. M., Zhang N., Buchanan S., Pilkey B., Coaker G., Iyer-Pascuzzi, A. S. (2025). *Tomato Roots Exhibit Development-Specific Responses to Bacterial-Derived Peptides. Plant Cell and Environment,* 48,8771-8787. https//doi:10.1111/pce.70164.

Ohgane, K., & Yoshioka, H. (2019). *Quantification of gel bands by an ImageJ macro, band/peak quantification tool.* Protocols.io. dx.doi.org/10.17504/protocols.io.7vghn3w.

Ron, M., Kajala, K., Pauluzzi, G., Wang, D., Reynoso, M. A., Zumstein, K., Garcha, J., Winte, S., Masson, H., Inagaki, S., Federici, F., Sinha, N., Deal, R. B., Bailey-Serres,

J., & Brady, S. M. (2014). Hairy root transformation using *Agrobacterium rhizogenes* as a tool for exploring cell type–specific gene expression and function using tomato as a model. *Plant Physiology, 166*, 455–469. https//doi:10.1104/pp.114.239392.

Schenk, S. T., & Schikora, A. (2015). Staining of callose depositions in root and leaf tissues. *Bio-protocol, 5*(6): e1429. https//doi:10.21769/BioProtoc.1429.

PART 2

HISTOLOGY ASSAYS

The protocols in this chapter cover root sectioning and staining for field and lab-grown plants.

2.1

HARVESTING ROOT SAMPLES IN THE FIELD

CONTRIBUTOR: DENISE CALDWELL

There are multiple types of fixative solutions, but when harvesting plant samples from the field we typically use formalin-aceto-alcohol (FAA) and then follow protocol 2.2 starting at the point in the steps where it says "Start here if using FAA fixation."

PRIMARY FIXATIVE SOLUTIONS FOR FIELD USE (MAKE FRESH EACH TIME)

FAA (50% ethanol, 10% formalin, 5% acetic acid):
Formalin is ~40% saturated solution of formaldehyde (can be purchased as 37% and is best within 6 months of purchase).

	100 ML	1 L
Ethanol (100%)	50 mL	500 mL
Glacial acetic acid	5 mL	50 mL
Formalin	10 mL	100 mL
ddH_2O	35 mL	350 mL

Store at 4°C.
For root harvest, fill a small glass vial with 30–50 ml of solution and keep on ice.
Harvest samples directly into the vial and keep on ice until returning to the lab.
Once in the lab, store samples at 4°C until ready to embed.

2.2

PARAFFIN EMBEDDING AND SECTIONING FOR LIGHT OR SCANNING ELECTRON MICROSCOPY

CONTRIBUTOR: DENISE CALDWELL

This protocol is adapted from Caldwell and Iyer-Pascuzzi (2019). For examples of paraffin embedding, sectioning, and imaging for either light microscopy or scanning electron microscopy, see Caldwell et al. (2017, 2024) and Caldwell and Iyer-Pascuzzi (2019).

MATERIALS

We purchase through Electron Microscopy Sciences (www.emsdiasum.com). Its catalog numbers are below, but you can purchase elsewhere if need be.

- Sodium cacodylate trihydrate (cacodylic acid, Sodium Salt, Trihydrate Sodium Cacodylate)
- Paraformaldehyde, EM Grade, Purified
- Sterilized distilled water (ddH$_2$O)
- Glutaraldehyde 25% solution, EM-grade distillation purified (example, www.emsdiasum.com catalog no. 16220)
- Ethanol 200 proof
- Paraffin (example, Paraplast X-tra Catalog number 19214)
- Tert-butyl alcohol (www.fishersci.com A401–500)

SOLUTIONS

Prepare all under chemical hood.

<u>0.2 M Cacodylate Buffer (pH 6.8) (Final Volume—200 ml)</u>

- 8.56 g sodium cacodylate (contains arsenic—avoid contact with skin)
- 200 ml ddH$_2$O
- pH to 6.8

<u>4.0% Formaldehyde in 0.1 M (pH 6.8) Cacodylate Buffer (Final Volume, 50 mL)</u>

- Under the chemical hood, heat 25 ml of ddH$_2$O to ~50–60° C in a small wide-mouth beaker with a small stir bar.
- Weigh out 2.0 g paraformaldehyde. Be careful not to breathe in fine powder.
  - 2.0 g paraformaldehyde
  - 25 ml H$_2$O
  - Add 10 M NaOH drops to the solution until dissolved
  - Cool in the hood
  - Mix 1:1 with 0.2 M cacodylate buffer (pH 6.8)

<u>3.0% Glutaraldehyde in 0.1 M Cacodylate Buffer (pH 6.8) (Final Volume, 50 ml)</u>

- 6 ml of 25% glutaraldehyde solution
- 19 ml ddH$_2$O
- 25 ml 0.2 M cacodylate buffer (pH 6.8)

<u>1.5% Glutaraldehyde/2% Paraformaldehyde–Primary Fixative Solution</u>

- Mix equal parts 3.0% glutaraldehyde with 4.0% formaldehyde
- Store in 4°C refrigerator
- Shelf life, 1 week

<u>0.1 M Cacodylate—Wash Buffer (pH 6.8)</u>

- Mix equal parts 0.2 M cacodylate buffer (pH 6.8) with ddH$_2$O
- Store in 4°C refrigerator
- Shelf life, 1 week

<u>25% Ethanol</u>

- 25 mL ethanol
- 75 mL ddH$_2$O

<u>50% Ethanol</u>

- 50 mL ethanol
- 50 mL ddH$_2$O

<u>TBA I</u>
- 40 mL ddH$_2$O
- 50 mL ethanol
- 10 mL tert-butyl alcohol*

<u>TBA II</u>
- 30 mL ddH$_2$O
- 50 mL ethanol
- 20 mL tert-butyl alcohol*

<u>TBA III</u>
- 15 mL ddH$_2$O
- 50 mL ethanol
- 35 mL tert-butyl alcohol*

<u>TBA IV</u>
- 45 mL ethanol
- 55 mL tert-butyl alcohol*

<u>TBA V</u>
- 25 mL ethanol
- 75 mL tert-butyl alcohol*

<u>Paraffin**</u>
- Paraplast wax melted to 54°C

General Notes:

*Tert-butyl alcohol is solid at room temperature above 25°C. A good way to keep it fluid is to keep it on top of the paraffin oven.

**The general rule for paraffin embedding is the slower the better! If you go through the dehydration and embedding steps too quickly, the tissue will collapse, and cell walls will be wavy instead of straight.

DO NOT use xylene in place of TBA.

FIXATION

Work in the chemical hood at all times!!!! Wear proper personal protective equipment at all times!

- Place 2–3 mL of either FAA or 1.5% glutaraldehyde/2% paraformalde-hyde–primary fixative solution into a glass scintillation vial (example, www.fishersci.com 03-341-73C). You can use smaller or larger vials, and you can use 15 ml Falcon tubes. It is very important that your sample be completely immersed in a fluid with 20–30 times more fluid than a sample. If you are already starting with samples fixed in FAA, skip to the dehydration step where it says "Start here if using FAA."

- Place ~2 mM sample in 1.5% glutaraldehyde/2% paraformaldehyde–primary fixative solution. For plants, samples should be placed into the fixative immediately after cutting from the plant.

- Place the scintillation vial without a lid into a vacuum desiccator (example, www.fishersci.com 08-594-15A). Apply low vacuum pressure for 2 hours. After 2 hours, remove samples from the vacuum desiccator by releasing vacuum pressure slowly.

- Remove the fixative with a fine-tip disposable transfer pipette (example, www.fishersci.com 13-711-26), being careful not to touch or pipette up the sample. Dispose of the spent fixative into properly labeled hazardous waste containers. Do not pour waste down a drain.

- Immediately replace the removed spent fixative with 2–3 mL 0.1 M cacodylate–wash buffer (pH 6.8). It is crucial that the sample is not susceptible to drying out. The drying-out process destroys cells by dehydration and causes problems later when trying to visualize the samples.

- Place the scintillation vial without a lid into a vacuum desiccator for 15 minutes, then release vacuum pressure slowly. Immediately remove spent 0.1 M cacodylate–wash buffer (pH 6.8) and replace with fresh 0.1 M cacodylate–wash buffer (pH 6.8). Repeat this step three times. Immediately continue to the dehydration step.

DEHYDRATION

- Start here if you are using FAA as the fixative. Remove FAA and replace it with 50% ethanol. Place the scintillation vial without a lid into a vacuum desiccator for 15 minutes and then release vacuum pressure slowly. Repeat this three times and then continue with the TBA I Step 3 bullet points below.

- Start here if you are following the protocol for 1.5% glutaraldehyde/2% paraformaldehyde as a fixative. Replace the removed spent 0.1 M cacodylate–wash buffer (pH 6.8) with 3 mL 25% ethanol solution. Place the scintillation vial without a lid into a vacuum desiccator for 15 minutes and then release vacuum pressure slowly.

- Remove spent 25% ethanol and replace with fresh 50% ethanol. Place the scintillation vial without a lid into a vacuum desiccator for 15 minutes and then release vacuum pressure slowly.
- Remove spent 50% ethanol and replace with TBA I. Place the scintillation vial without a lid into a vacuum desiccator for 15 minutes and then release vacuum pressure slowly.
- Remove spent TBA I and replace it with TBA II. Place the scintillation vial without a lid into a vacuum desiccator for 15 minutes and then release vacuum pressure slowly.
- Remove spent TBA II and replace it with TBA III. Place the scintillation vial without a lid into a vacuum desiccator for 15 minutes and then release vacuum pressure slowly.
- Remove spent TBA III and replace it with TBA IV. Place the scintillation vial without a lid into a vacuum desiccator for 15 minutes and then release vacuum pressure slowly.
- Remove spent TBA IV and replace with TBA V. Place the scintillation vial without a lid into a vacuum desiccator for 15 minutes and then release vacuum pressure slowly.
- Remove spent TBA V and replace with 100% tert-butyl alcohol. Replace the cap securely—do not let it evaporate. Place the scintillation vial on top of the paraffin oven or another warm place (incubator at 26–28°C is okay). Leave for 24 hours.
- After 24 hours, remove spent tert-butyl alcohol and replace it with fresh tert-butyl alcohol. Replace the cap securely! Place the scintillation vial on top of the paraffin oven or another warm place (incubator at 26–28°C is okay). Leave for 24 hours.
- Repeat this a minimum of three times. It is essential that all water be removed from the sample before proceeding to the next step and that the lid be secure. This is a good stopping point if you have to process many samples (example: during field season). Samples can be kept in tert-butyl alcohol for weeks as long as they are NOT exposed to air.

INFILTRATION

- Remove spent tert-butyl alcohol and replace it with fresh tert-butyl alcohol. Always make sure that the samples are covered with alcohol.

- Fill the remainder of the scintillation vial with melted paraffin wax. Replace the cap securely. Place the scintillation vial into the paraffin oven (or incubator at ~54°C) for 24 hours with a cap on. After 24 hours, remove the cap from the scintillation vial and place it back into the paraffin oven.
- Leave the vial in the oven for 24 hours to allow the tert-butyl alcohol to evaporate. You will notice that the amount of liquid has reduced, but the sample should remain covered with paraffin.
- Carefully pour out paraffin/tert-butyl alcohol and replace with freshly melted paraffin. Leave the cap off and place the vial back into the paraffin oven for 24 hours. Repeat this step a minimum of three times.

EMBEDDING FOR LIGHT MICROSCOPE VIEWING

- Warm a glass petri dish on a 100°C heating plate and exchange the paraffin in the vial with fresh paraffin. Pour into the warm petri dish using a wood applicator or toothpick with a flat base, and use the applicator to move the fluid around the sample(s).
- Fill a peel-away disposable embedding mold with fresh paraffin wax and scoop the sample into the mold so that it stands perpendicular in the mold (use the wood applicator or be very gentle with forceps). Allow the block to cool at room temperature for 24 hours.

SECTION SAMPLES (PROTOCOL IS FOR LIGHT MICROSCOPY)

Mounting

- Remove the paraffin block from the peel-away mold and attach the base of the block onto Tissue-Tek embedding rings.
- Melt around the base of the block so that it is solidly attached to the ring and allow to cool for 24 hours.
- Trim the block with a razor blade to form a trapezoid on the block surface around the sample.

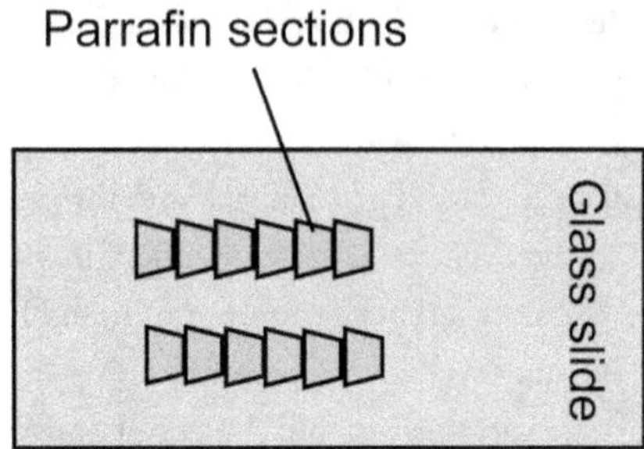

FIGURE 2.2.1. Example of sectioned samples on a glass slide

Sectioning

- Cut the blocks until a smooth surface is obtained and then cut to appropriate thickness (7 to 12 μm generally for light microscopy, but you will have to optimize what works for your samples) using a microtome.
- Place water on a glass slide sitting on a slide warmer at 40°C. Place the cut ribbon onto the water on the slide. After 5 minutes, remove excess water with a Whatman filter (place next to the ribbon and allow water to absorb). Allow slide to dry for 24 hour on the slide warmer. Sections should be on the slide as in Figure 2.2.1.

DEPARAFFINIZATION AND STAINING FOR LIGHT MICROSCOPY

- After you have sectioned your samples and they are on the glass slide, place the slide into the glass staining carriage.
- Set up your deparaffinization and staining jars as shown in Figure 2.2.2. Stains can vary—we usually start with a polychromatic dye called toluidine blue. Other common histological stains can be found in the literature.
- The first jar will contain 100% xylene. Place the carriage in the first staining jar "Start" for 5 minutes and then move one space to the next staining jar.

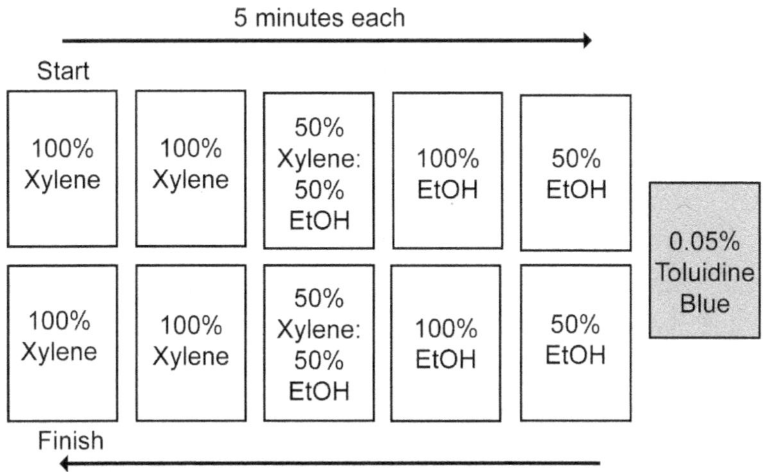

FIGURE 2.2.2. Deparaffinization and staining setup

- After the carriage has moved through all the stain jars, remove a slide from the "Finish" stain jar, apply 3 drops of Permount to the surface of the glass slide, and place a 22 x 50 mM glass coverslip on the surface. Lay the glass slide on a flat surface to dry.

SECTIONING AND DEPARAFFINIZATION FOR SCANNING ELECTRON MICROSCOPE

- Section samples from paraffin blocks at 20 μm.
- Tape a round cover glass to a glass slide with double-sided tape. It is best to tape down just a little part of one side; otherwise, removing the tape from the glass slide without breaking the cover glass is difficult. See Figure 2.2.3.
- Place a drop of water on the cover glass and float paraffin samples on it. Remove excess water and allow it to dry for 24 hours.
- Fill three stain jars with 100% xylene. Place the glass slide sample into xylene for 5 minutes. After 5 minutes, remove the slide and place it into a new stain jar filled with fresh xylene. Repeat this a total of three times.
- Place glass slide into 100% ethanol for 5 minutes.
- Remove and allow to dry.

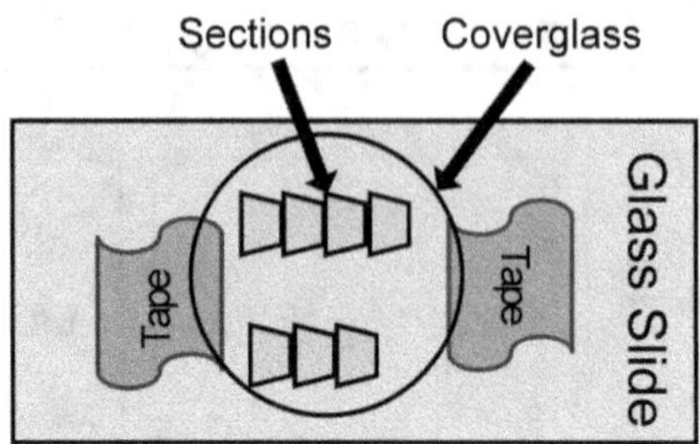

FIGURE 2.2.3. Example of slide with sections for scanning electron microscopy

MOUNTING SAMPLES

- Attach double-sided tape to the proper specimen mount required for the SEM that you are using. We use a simple 3.1 mM tapered end pin with a 12.7 mM slotted head (EMS #75200), but this should be determined by your SEM manager.
- Remove the deparaffinized cover glass and attach it to the specimen mount.
- To reduce the charging of the sample, paint a line from the pin of the specimen holder to the cover glass with a silver conductive coating (EMS #12684–15) or whatever is recommended by the SEM manager.
- Sputter coat for 90 seconds.
- Place in a dust-free environment until they are ready to be viewed in the microscope.
- Note that if you are going to operate the SEM yourself, you will need to take a training course.

2.3

ROOT EMBEDDING, VIBRATOME, AND STAINING PROTOCOL

CONTRIBUTOR: SANA DOTY

This protocol was adapted from Ron et al. (2013). There are multiple steps: embedding tissue, rehydrating plugs, sectioning plugs, and staining and imaging sections.

EMBEDDING TISSUE IN PLUGS

MATERIALS NEEDED

100mL 2% agarose solution

Double-sided tape

Plug mold (cut top of 15 mL Falcon tube or 1.5 mL Eppendorf)

2 forceps

Plant roots of interest

Razor blades

Autoclaved water or FAA (see recipe in Chapter 3.1)

SET UP PLUG MOLD

1. Place double-sided tape down onto benchtop and position mold onto tape with flat side down (the uncut side).

2. Cut desired root segment off of plant with a razor blade. Make the segment approximately the length of your mold (Example: A cut top off a 15 mL falcon tube is ~13 mM tall).

3. Pour hot-warm agarose into plug. *Note: It will be viscous, so be careful not to overpour.*

4. Position root segment(s) in agarose with your forceps. Plugs can hold up to 3–4 root segments. Use both forceps to hold the root segment in place so it stays straight. Either wait until agarose is slightly cooler and push root straight or have the top of the segment lean against the forceps resting on top of the mold to help it dry straight.

Note: Do not move roots when agarose is thickening/cooling. The vibrations from movement will cause there to be airspace around root, and the root will fall out when sectioned.

5. Repeat with all root segments, keeping the treatments/genotypes in separate plugs.

6. Once dry, pop plugs into a clean 20 mL scintillation vial and label. If using the day of or 1–2 days after making, fill the vial with sterile water to keep the plugs hydrated. If using 2+ days after making, fill vials with cold FAA to fix plugs until further processing.

7. Store vials in 4°C.

REHYDRATING PLUGS FROM FAA

MATERIALS NEEDED

Plugs in vial of FAA (see 2.1 for recipe)
Transfer pipettes
70%, 50%, 30%, and 10% ethanol
Autoclaved water
Waste bucket labeled per REM protocol

1. In a chemical fume hood, carefully remove FAA from vials with a transfer pipette. Discard FAA in labeled waste beaker, and discard pipette in hazardous waste.

2. Add 70% EtOH until all plugs are covered. Let sit for 30 min.

3. Remove 70% EtOH with a transfer pipette and discard in waste beaker/hazardous waste.

4. Repeat Step 2–3 with 50% EtOH, 30% EtOH, and then 10% EtOH.

5. After rehydrating samples, leave samples submerged in sterile water and

store in 4°C until processed. Samples can be left in water for 1–2 days. If leaving longer than that, refix in FAA.

SECTIONING PLUGS WITH A VIBRATOME

MATERIALS NEEDED

Vibratome	6-well plate
Rehydrated plugs	Autoclaved water
Ice bucket w/ice	Super glue
Paint brush	Waste beaker
Forceps	Double-edged razor blades
Microscope slides + covers	Light microscope

Turn on Vibratome (Our Model: Pelco EasiSlicer).

Set amplitude = 10

Set speed = 4

Set light = high

Set pause = off

Set return = Auto

PREPARE PLUG FOR SECTIONING

1. Take the plug out of the vial with forceps or pour it out into a beaker. Cut off a vertical section from the plug where there are no roots present (see image). The plug should look like a "D" (Figure 2.3.1a).
2. Cut off a horizontal section off the top (uneven) part of the plug (just a few mM) (Figure 2.3.1b).

a.  b. 

FIGURE 2.3.1. Example of plug shapes for vibratome sectioning

3. Glue bottom of the plug down onto the vibratome stage (black rectangle) with a little dot of superglue. Let dry completely.
4. Position the stage in the blue compartment (there are magnets below it that line up with the compartment).
5. Fill the blue compartment with ice, and fill the stage with ice-cold water. Replace ice in the blue compartment as it melts. Plugs cut better when cold.

SET UP VIBRATOME BLADE

1. With the L-wrench, remove the screws on the top of the blade holder.
2. Insert a fresh 1/5 of a double-sided blade.
3. Put the blade holder back on and tighten the screws.
4. Secure onto vibratome with the large screw on top of the blade holder.

SET UP CUTTING WINDOW

1. Set the speed = 10, pause = on, and return = manual.
2. Reverse the blade all the way back.
3. Inch the blade forward until it lines up with the front edge of the plug. Make sure the blade is positioned over the plug so it doesn't accidentally cut. (Turn the black dial counterclockwise to move it upward).
4. Once in position, unscrew the dial on the back-left side of the vibratome and move it toward you until it stops. Tighten the dial there. This is your starting position.
5. Move the blade forward until it stops. Inch it backward until is stops right after the back end of the plug. Once in position, unscrew the dial on the front-left side of the vibratome and move it away from you until it stops. Tighten the dial there. This is your ending position.
6. Move the blade back and forth to double-check the positions. Adjust dials as needed.
7. Once the cutting window is set, return settings to Step 1 numbers. Rotate the black dial clockwise to bring down the blade. Position the blade just at the top of the plug. Switch it to "forward" to begin cutting. If the blade does not cut anything the first time, keep moving the blade down until it starts cutting.
8. Discard sections that are not full cuts. You can collect the sections as they finish with the paint brush.

9. When the first full cut happens or the first full cut that has your root sections present, turn the dial only for the desired thickness.

Note: On this model 80 microns cuts pretty well, but you may have to adjust for other models/needs.

10. After cutting the first few sections, examine the sections on a light microscope to check thickness and root straightness.
11. Once the desired thickness and straightness have been achieved, you may cut your sections without checking them.
12. Collect sections in a labeled 6-well plate with water in the wells. Once the desired number of sections have been made, move the blade upward. Remove the blade and place in the holder. Discard the water in the stage and scrape off plug/superglue using a razor blade. Dry-stage before setting up the next plug.

CLEANUP

1. Dump all water and ice from the stage/compartment. Dry with paper towels.
2. Remove the blade from the machine and take out the razor. Discard in sharps container.
3. Return unused portions of plugs to a vial and fix with FAA for long-term storage. Store at 4°C.
4. Store 6-well plate with sections in 4°C until staining. Process within 24 hours.

STAINING AND IMAGING SECTIONS

Here we use phloroglucinol staining for lignin, but other stains can be used.

MATERIALS

Sections	Phloroglucinol
Microscope slide + cover slips	HCL (full strength)
Kimwipes	Transfer pipettes
2x 15 mL Falcon tubes	Timer
Sterile water	Light microscope

MAKE PHLOROGLUCINOL STAIN (NOTE: MAKE FRESH THE DAY OF STAINING)

1. Add 4 mL of sterile water to a 50 mL Falcon tube.
2. Slowly add 1 mL of full-strength HCL in the fume hood. An exothermic reaction will occur, so adding slowly will prevent it from boiling.
3. Add in 0.1 g of phloroglucinol. Vortex to mix. Solution should be saturated, so there should be a little bit of undissolved stain.
4. Filter-sterilize into a 15 mL Falcon tube covered in foil (stain is light-sensitive).

TO IMAGE:

1. Turn on the microscope and the camera. Open the camera app. We use an Olympus Bx53 equipped with a SPOT Idea 5.0 Mp color digital camera.
2. Set your image preferences and save. *Every image taken should use the same settings.*
3. Using a soft brush, carefully transfer a section to a microscope slide, avoiding touching the middle of the section where the root is.
4. Add a drop of the phloroglucinol stain. Let sit for 1 min before covering with a cover slip to allow the stain to oxidize. Wait 8–10 minutes before imaging for the stain to penetrate. Image will look like Figure 2.3.2). **Do not image past 12 minutes, as the stain will start to degrade the tissue.**
5. Repeat with all sections.

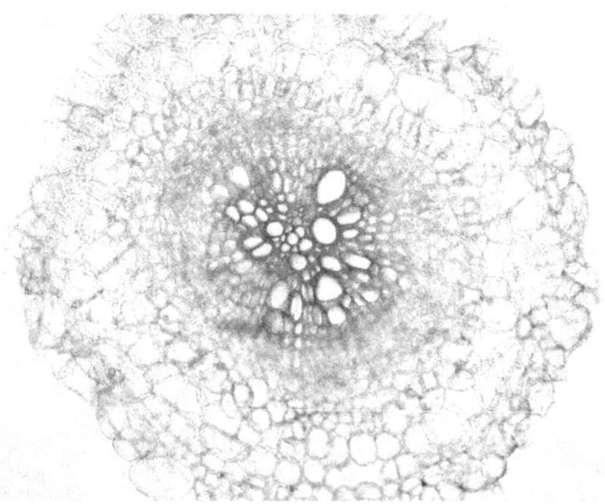

FIGURE 2.3.2. Example of a cut and phloroglucinol stained root. Section was imaged with light microscopy

WORKS CITED

Caldwell, D., Kim, B. S., & Iyer-Pascuzzi, A. S. (2017). *Ralstonia solanacearum* differentially colonizes roots of resistant and susceptible tomato plants. *Phytopathology, 107*, 528–536. https//doi:10.1094/PHYTO-09-16-0353-R.

Caldwell, D., & Iyer-Pascuzzi, A. S. (2019). A scanning electron microscopy technique for viewing plant-microbe interactions at tissue and cell-type resolution. *Phytopathology, 109*, 1302–1311. https//doi:10.1094/PHYTO-07-18-0216-R.

Caldwell, D. L, da Silva, C. R., McCoy, A. G., Avila, H., Bonkowski, J. C., Chilvers, M. I., Helm, M., Telenko, D. E. P., & Iyer-Pascuzzi, A. S. (2024). Uncovering the infection strategy of *Phyllachora maydis* during maize colonization: A comprehensive analysis. *Phytopathology, 114*, 1075–1087. https//doi:10.1094/PHYTO-08-23-0298-KC.

Ron, M., Dorrity, M. W., de Lucas, M., Toal, T., Hernandez, R. I., Little, S. A., Maloof, J. N., Kliebenstein, D. J., & Brady, S. M. (2013). Identification of novel loci regulating interspecific variation in root morphology and cellular development in tomato. *Plant Physiology, 162*, 755–768. https//doi:10.1104/pp.113.217802.

PART 3

FUNCTIONAL ASSAYS FOR ROOT-ASSOCIATED BACTERIA

These assays are good for initial functional characterization of root-associated bacteria before moving on to more in-depth analysis. See Tran et al. (2022) examples and more details.

3.1

BLEACH STERILIZATION OF PLANT ROOT TISSUE AND DNA EXTRACTION FROM ROOT ENDOSPHERE

CONTRIBUTOR: SANA DOTY

This protocol is used to extract DNA from the root endosphere. After extraction, DNA can be used for library generation for microbiome analysis.

MATERIALS

Plant root tissue (frozen or fresh)
Sterile water
5% bleach (freshly made)
   Shaker

Forceps
15 or 50 mL conical tubes (depending on tissue size)
Liquid nitrogen

PROTOCOL

1. Rinse root tissue in sterile water (or sterile buffer) until roots are cleaned of any soil debris.
   a. Either gently brush roots with a soft-bristle paint brush under water or place roots in a conical tube of water and gently shake to remove debris.

2. Surface-sterilize root tissue by placing in a conical tube of freshly made 5% bleach and gently shake in a shaker (or invert it gently if a shaker is unavailable) for 2 minutes.
   a. Note: Be mindful of bleach percentage and exposure time. Making the bleach fresh will ensure that you are not using an old bleach solution that has evaporated and increased concentration.
3. Immediately following the 2-minute sterilization, rinse roots in sterile water at least three times to remove all bleach residue.
4. Freeze tissue in a conical tube or foil envelope using liquid nitrogen.
5. Store at -80°C until further processing or move on directly to DNA extraction.

DNA Extraction from Liquid Culture

NOTES

- Place a 1.5 mL tube of nuclease-free water in the water bath prior to starting the protocol.
- This protocol is modified to use the Norgen Soil DNA Isolation Kit for bacterial gDNA extraction from liquid culture.
- Spike LB broth with bacterial colony 24 hours prior to DNA extraction and leave in shaker in incubator overnight at 37°C.

PROTOCOL

1. Remove bacterial liquid cultures from the incubator shaker.
2. Place cultures in A big centrifuge and centrifuge at max speed for 5 minutes, until the bacteria has pelleted at the bottom and the supernatant is clear.
   a. If the supernatant is not clear after 5 minutes, centrifuge again for additional time until all of the bacteria has pelleted.
3. Pour off supernatant carefully into your waste beaker without disturbing the pellet.
4. Add 750 µL of Lysis Buffer G to snap tube and resuspend pellet by vortexing the solution until the entire pellet has been resuspended.

5. Add 200 μL of Lysis Additive A and vortex to mix for at least 30 seconds.

6. Centrifuge tubes at max speed for 5 minutes.

7. Transfer clean supernatant (via pipette) to a labeled 1.5 mL tube without disturbing the pellet.

8. Add 100 μL of Binding Buffer I and mix by inverting tube a few times. Incubate on ice for 5 minutes.

9. Centrifuge tubes at max speed for 2 minutes.

10. Carefully transfer all of the supernatant into a clean 1.5 mL tube without disturbing the pellet.

11. Add 50 μL of OSR Solution and mix by inverting. Incubate on ice for 5 minutes.

12. Centrifuge tubes at max speed for 2 minutes.

13. Carefully transfer all of the supernatant into a clean 1.5 mL tube without disturbing the pellet.

14. Add 400 μL of Lysis Buffer QP and 550 μL of 100% ethanol to the tubes. Vortex to mix.

    a. If this volume of solution cannot fit into the tubes, split the sample into two tubes.

15. Add 600 μL of this mixed lysate mixture into a spin column assembled within a collection tube.

    a. It is okay if there is lysate left over. You will repeat this again after centrifuging.

    b. If a sample was placed into two tubes for step 14, then recombine them into one spin column when repeating this step.

16. Centrifuge tubes at max speed for 30 seconds. Discard flow-through in a waste container.

17. Repeat step 15 and 16 until all of the lysate from the tubes has gone through the spin column.

18. Add 500 μL of wash solution A to the spin column and centrifuge at max speed for 1 minute.

19. Discard flow-through and repeat step 18.

20. Centrifuge columns at max speed for 2 minutes to ensure that all of the wash solution has been removed.

21. Place the spin column in a clean labeled 1.5 mL tube.

22. Add 50 μL of the warmed nuclease-free water from the water bath.

    a. Carefully pipette water in the center of the spin column resin.

23. Incubate at room temperature for 2 minutes.

24. Centrifuge tubes at max speed for 2 minutes.

25. Measure DNA concentration of samples using the Nanodrop and record on the tube and in a lab notebook.

26. Place samples in the -20°C freezer for long-term storage.

3.2

PATHOGENIC BACTERIA ROOT COLONIZATION ASSAYS IN PLATES

CONTRIBUTOR: ABIGAIL ROGERS

MATERIALS

This protocol uses part of Chapter 3.3, so also check the materials needed in that protocol.

• Forceps	• Lab notebook + pen
• Sterile water	• Sharpie
• P1000, P200, P20 + tips	• Paper towels
• Eppendorf tubes	• 70% ethanol spray
• Water agar plates	• Multichannel pipette + yellow tips
• Tomato seedlings	• CPG agar plates with TZC*

Note: Prior to performing experiments, a preliminary study should be conducted to determine what OD600 is equivalent to 10^8 CFU/ml and 10^5 CFU/mL, as this may vary between bacterial strains.

PROTOCOL

Day -6: Sterilize and stratify enough tomato seedlings to plate 8 seedlings on each water/agar plate.

Although 8 seeds will be planted, only 3 seedlings will be harvested and assayed. One seedling will be removed on inoculation day. Planting extra seeds will help prevent loss of samples due to poor germination.

Day -5: Plate tomato seeds on 1% water/agar plates (any size plates; see water agar recipe in Chapter 3.1). You can perform up to two treatments per plate, so four seeds should be plated to the left and the right of the middle of the plate to separate treatments. Secure the plate lids with micropore tape and move to light. Allow the seeds to germinate and grow along the surface of the media until inoculation day. This may take anywhere from 4–5 days to grow depending on the time of year and lighting conditions.

Day -2: Plate *Ralstonia* on CPG + TZC and grow at 30°C for two to three days.

Day 0: Prepare *Ralstonia* inoculum and inoculate roots in the laminar flow hood.

> Make a solution at OD600 = 0.2, equivalent to 10^8 CFU/ml, and dilute the inoculum to 10^5 via serial dilutions. Remove tomato seedlings until there are a total of six seedlings per plate (three seedlings per treatment). Pipette 50 μL of inoculum onto the root tip of the seedlings and let the inoculum dry.

> While inoculum is drying on roots, plate serial dilutions of inoculum on CPG +TZC plates from 10^{-1} to 10^{-8} to calculate CFU/mL of initial inoculum.

> Following inoculation, secure plate lids with micropore tape and move the plates back to the light rack to be assayed at 24, 48 and 72 hours postinoculation.

> At 24, 48 and 72 hours postinoculation, each day select 3 roots from each treatment from each plate to pool for harvest and follow the protocol in Chapter 3.3 starting at step 5 (griding roots).

3.3

PATHOGENIC BACTERIA ROOT COLONIZATION ASSAYS IN SOIL

CONTRIBUTOR: SANA DOTY

We use this assay to assess colonization of *Ralstonia solanacearum* in roots, but it would work with any bacterium.

MATERIALS

• Weigh scale	• Autoclave bags
• Plastic cutting board	• 5 gallon bucket
• Thick-end paint brush	• Metal sieve
• Forceps	• Lab notebook + pen
• Autoclaved mortar and pestles	• Sharpie
• 2x 4 L plastic beakers	• Paper towels
• 1x small autoclave tub	• 70% ethanol spray
• Soaking paper	• 100% ethanol
• Sterile water	• Ethanol burner
• P1000 + tips	• Empty petri dish
• 15 mL conical tubes	• Multichannel pipette + yellow tips*
• Weigh scale	• 96-well plates*
• Empty flat/container for soil	• P20 + tips*
• Razor blades (single-sided)	• P200 + tips*
	• CPG agar plates with TZC*

*Only needed when diluting and plating in the bacterial hood.

PROTOCOL

1. Set up root washing station:
   a. Lay down soaking paper.
   b. Fill the two 4 L plastic beakers with distilled water (from sink tap). One will be for the dirty roots, **beaker 1**. The other will be for slightly cleaner roots, **beaker 2**.
   c. Fill a small autoclave tub with distilled water, ~0.5–1 in in thickness.
   d. Place an empty flat next to this setup to collect the discarded soil. Soil will later be put into an autoclave bag for biohazard waste processing.
   e. <u>Note</u>: If there are no tables available, this setup should be done on the bottom level of a big cart, leaving the top level for the grinding setup.
2. Clean soil from roots:
   a. Carefully remove plants from their pots.
   b. Over the flat, gently massage roots to remove large clumps of soil.
   c. <u>Tips</u>:
      Using your fingertips in a cupped form while moving the roots in and out of your cupped fingers is effective at removing the soil clumped around the primary root.

      If you do not need the shoot, cutting it off at the cotyledons allows for easy cleaning and a little hypocotyl "handle" as you continue the next steps.
   d. Once most of the soil has been removed, dunk the roots in **beaker 1** a few times before dunking the roots again in **beaker 2**. By this point, most of the soil should be removed except for soil particles close to the root epidermis.
3. Gently brush roots with the wide-edged paint brush in the autoclave tub of water to remove the finer soil particles.
   a. <u>Tip</u>: Moving the brush in a side-to-side motion perpendicular to the primary root is effective for removing the soil lodged in between the lateral root emergence zones.
   b. This step will require more pressure if using larger roots. For smaller roots, be very gentle at this step to minimize the damage to the roots.
4. Once the roots are adequately clean, cut the root off from any remaining shoot at the root-shoot junction using a razor blade.
   a. <u>Note</u>: Doing this on the plastic cutting board lined with paper towels helps remove excess water from the roots and gives you a convenient

method to carry your roots from the washing station to your grinding station.

5. Grinding roots: Measure and record the weight of the root that you will be grinding.

6. Using forceps, take the root and dip it in 100% ethanol stored in a petri dish or some other secondary container.

   a. Tip: Gathering the root into a loose clump before grabbing with the forceps will help with this step and the next step.

7. Dab roots with a paper towel to remove excess ethanol.

   a. Note: For larger roots, squeeze roots gently between two paper towels to remove excess ethanol.

8. Pass roots quickly through the flame of an ethanol burner or the Bunsen burner. Let the ethanol burn off of the root surface.

   a. Note: For larger root clumps, pass the roots through the flame a few times until no more ethanol remains. Ethanol tends to stay within the root clump and could prevent *Ralstonia* growth when plated if not burned off entirely.

9. Once all the ethanol is burned off of the surface of the root, dry-grind the root in a sterile mortar.

   a. Tip: Lightly pounding thicker roots with pestle allows for easier grinding.

10. Add 1 mL of sterile water to the mortar and continue grinding until you've reached a paste-like consistency.

11. Transfer material from the mortar into a labeled tube.

   a. Note: For smaller roots, use 1.5 mL tubes. For larger roots, use 15 mL conical tubes.

   b. Tip: Use the pestle and your forceps to help transfer all the material from the mortar into the tube.

12. Wipe the mortar and pestle clean with ethanol and paper towels before repeating steps 5–11 on the next sample.

   a. Note: You can reuse a mortar and pestle with samples from the same experimental group with a simple wipe between samples. However, a fresh and sterile mortar and pestle should be used between experimental groups.

   b. Tip: If you do not have enough autoclaved mortar and pestles for each experimental group, then you can ethanol-sterilize them between samples. Spray ethanol into the mortar and pestle after it has been wiped clean. Then, light the mortar with a lighter or forceps that have been dipped and lit with ethanol. Once it is burned off and cooled, this mortar and pestle is now sterile and can be used again.

13. Once all the roots have been processed, you are ready to dilute and plate for the colonization assay.

14. Cleanup: Throw all disposable waste into an autoclave bag, including the soil.

15. Ethanol-spray and wipe anything that cannot be autoclaved.

16. Dump all water through a metal sieve into a 5-gallon bucket. This sieve will catch any soil particles, which should be discarded with the soil. Pour some bleach into the bucket, and after 24 hour dump bleached water into a sink.

17. *Plating for a colonization assay* (perform in the sterile hood): Label a 96-well plate with the appropriate sample names at the top and -1 to -8 going down the rows.

18. Using a multichannel pipette, fill the wells with 180 µL of sterile water.
    a. <u>Note</u>: The multichannel pipette works best with the yellow p200 tips.
    b. <u>Tip</u>: use the lid of the 96-well plate as your water trough. Any unused water can be discarded, and the lid can be wiped dry with a kimwipe.

19. Add 20 µl of the samples to the "-1" well of the appropriate column. Do this for all of the samples before continuing to the next step.

20. Using the multichannel pipette, mix the "-1" dilution row by pipetting up and down, and then take 20 µL of this dilution and mix into the "-2" rows. Continue this until all of the dilutions have been completed.

21. Label CPG agar plates (with TZC and optionally tet10 if working with *Rs*K60-GFP) with the sample names as columns and dilutions as rows.
    a. <u>Note</u>: Keep experimental groups on separate plates, organized by genotype, treatment, and dpi.

22. Starting with the most diluted row (-8 most commonly), plate 20 µL of dilution in a droplet form in the appropriate grid of the CPG plate. Then, continue this with higher dilutions until you've plated all 8.

23. Let plates air-dry in the hood before incubating at 28°C for 48–72 hours.
    a. <u>Tip</u>: If space on the plate allows, swirl plate in a circle to expand droplet surface area. This will make the droplets dry quicker.

24. When *Ralstonia* colonies have grown, record how many colonies have grown at the second most dilute grid they grew on and at what dilution you counted from.

25. Calculate CFU/mL of *Ralstonia* using this formula:

CFU/ml = Number of Colonies/ (0.02)(10x),

where 0.02 represents how much of the solution was plated in mL (20 μL = 0.02 mL), so adjust accordingly, and X = dilution you counted from (ex. "-3").

26. Calculate CFU/g of root tissue using this formula:

CFU/g = [CFU/ml]/grams of root tissue

3.4

ISOLATING BACTERIA FROM THE RHIZOSPHERE AND ROOT ENDOSPHERE

CONTRIBUTOR: SANA DOTY

MATERIALS

Plant tissue samples
LB agar plates
(Optional: Other selective medias)
Sterile water
70% EtOH spray
Kimwipes
Fine-tipped Sharpie
Scale

Mortars and pestles
Forceps
Glass spreader
P1000 and P200 pipettes + tips
Waste beaker
1.5 mL Eppendorf tubes
1x PBS buffer
28°C incubator

PROTOCOL

Preparation

1. Label the LB agar plates (and other media such as KB if you have it) with the date, sample ID, sample type, and a three-row grid. There should be 2 plates/sample for each plate type, containing 6 different dilutions (0, -1, -2, -3, -4, -5), three on each plate.

    a. Example plate label:

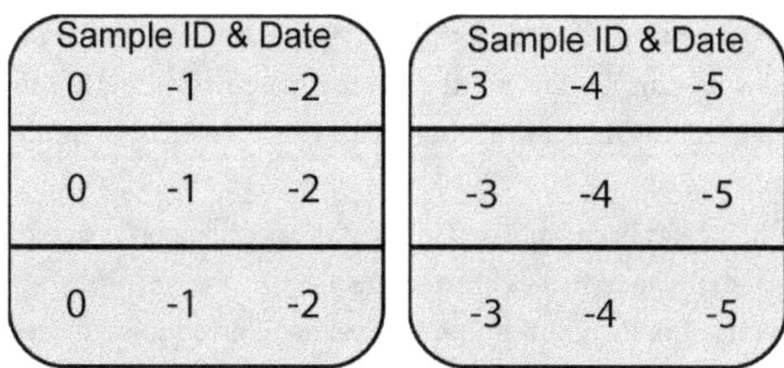

FIGURE 3.4.1. Plating scheme for dilution series

2. Label six 1.5 mL Eppendorf tubes from 0 to -5 and the sample ID/type. There should be one set of six/sample.
3. In a hood, add 900 µL of 1x PBS buffer to tubes -1 to -5.

Processing Samples

FOR PLANT TISSUE SAMPLES:

1. (Optional) If epiphytes from leaves need to be collected, place plant tissue in a 15 mL conical tube with 3 mL of 1x PBS buffer and vortex aggressively to shake epiphytes off of the plant tissue. This solution is the EPIPHYTE sample. Pipette 1 mL of the EPIPHYTE sample into a tube labeled with "0" and the sample ID/type. The "0" indicates nondiluted sample.
2. If the RHIZOSPHERE sample will be collected, shake rhizosphere soil from roots into a 15 ml conical tube with 3 mL of 1x PBS buffer and vortex aggressively. Note that you may need to use more PBS if you have more roots. Pipette 1 mL of the RHIZOSPHERE sample into a tube labeled with "0" and the sample ID/type. The "0" indicates a nondiluted sample.
3. If epiphytes are not going to be collected, place in 100% EtoH using forceps.
   a. For leaves/floral tissue, let soak for 30 seconds.
   b. For roots, just fully submerge in EtOH and remove.
3. Remove plant tissue from EtOH.
   i. For leaves: Rinse leaves in fresh sterile water. This sample has now been surface-sterilized and is ready for grinding. Water should be changed between samples.

    ii. For roots: Dab off excess EtOH on a paper towel/kimwipe and quickly pass the sample through the flame of the EtOH burner using forceps. The EtOH on the root surface will burn off, leaving the surface sterilized.

4. Once tissue has been sterilized, place it in a sterile mortar and pestle and dry-grind until a paste has been formed.

5. Add 1 mL of 1x PBS buffer to the mortar and grind some more until the solution is thoroughly ground.

6. Pipette or scrape the solution into the tube labeled with "0" and the sample ID/type. This is the ENDOPHYTE sample.

   NOTE: Use EtOH to sterilize forceps between samples.

   NOTE: If the liquid is cloudy, centrifuge briefly to pellet any soil particles before pipetting.

7. Take 100 µL of the "0" solution and pipette it into the "-1" tube. This is the first 1:10 dilution. Mix the tube by pipetting up and down or vortex the tube.

8. Repeat step 5, with 100 µL of "-1" solution into "-2" tube. Continue to repeat until all the dilutions have been made.

Plating Dilutions

1. Pipette 200 µL of dilution -5 onto the appropriate plate in the appropriate row.

2. Repeat step 7 with all the dilutions for that sample. The same pipette tip can be used if done in ascending order (-5 > 0).

3. Once all of the dilutions have been plated, spread solution using a sterilized glass spreader. BE SURE to keep the solution within the row it belongs in. Do your best to avoid cross-contamination between dilutions. The same spreader can be used if done in ascending order (-5 → 0). Be sure to EtOH sterilize the spreader between samples.

4. Let plates dry in the hood before placing in the 28°C incubator.

5. Let plates incubate for 24–48 hours.

Isolating Colonies

1. Label an LB agar plate with a 6 x 6 grid. Label each square with a number from 1 to 36. If there are more than 36 individual colonies for each sample type, make a second grid plate and label the squares 37 to 72.

2. Identify individual colonies on dilution plates and colonies of interest. You can denote these colonies by marking the back of the plates with a Sharpie dot. Preferentially choose colonies from the most diluted rows.

3. Using a sterile P200 pipette tip, carefully scrape some of the individual colony and streak a line onto the grid plate in one square, starting with square 1.

4. Repeat step 4 on all denoted colonies from dilution plates, streaking onto a different square for each individual colony. Take note of which grid numbers are associated with which sample ID.

    Note: Colonies from different sample IDs can be combined on one grid plate if they are the same sample type. (Example: All root endophyte dilution plates can be on one grid even if they are different individual plants.)

5. Put grid plates in a 28°C incubator overnight. This will be the MASTER PLATE of your ISOLATES. From here, you can test each of your isolates for different functions such as auxin production, siderophore production, and phosphate solubilization (described in Chapters 3.5, 3.6, 3.7).

3.5

AUXIN PRODUCTION FROM BACTERIA

CONTRIBUTORS: TRI TRAN AND SANA DOTY

Note: This is a good, fairly quick protocol to estimate auxin production from bacteria, but it is not extremely sensitive. It is adapted from Gordon and Weber (1951).

MATERIALS

YPM broth with 0.1% tryptophan
Salkowski reagent (fresh every time)
(0.5 M FeCl3, 1.35 mg FeCl3 in
  10 mL ddH$_2$O)
(70% Perchloric Acid) *CORROSIVE*
IAA standards (acetone, IAA)
Drying oven
Spectrophotometer
70% EtOH + kimwipes for cleaning

14 mL snap tubes for liquid cultures
200 mL glass bottle (for
  Salkowski reagent)
20 mL glass pipette (for
  Salkowski reagent)
Glass test tubes
1.5 mL Eppendorf tubes
P1000 + tips
28°C incubator + shaker
Glass waste beaker

PROTOCOL

Day 1: Preparation of Liquid Cultures

1. Label the 14 mL snap tubes with each isolate name.
2. Add 4 mL of YPM broth + 0.1% tryptophan into the tubes.

3. Inoculate the tubes with a single colony using a p200 tip.

4. Incubate tubes in the 28°C incubator on the shaker for 4 days.

Day 5: Experiment Day

1. <u>Prepare IAA Standards</u>
    a. 1000 µg/mL stock: Add 10 mg of IAA to 10 mL of acetone in a sterile 15 mL conical tube.
    b. 100 µg/mL stock: Add 1 mL of the 1000 µg/mL to 9 mL YPM+0.1% tryptophan broth in a sterile 15 mL conical tube.
    c. 50 µg/mL stock: Add 500 µL of the 1000 µg/mL to 9.5 mL YPM+0.1% tryptophan broth in a sterile 15 mL conical tube.
    d. 20 µg/mL stock: Add 2 mL of the 100 µg/mL to 8 mL YPM+0.1% tryptophan broth in a sterile 15 mL conical tube.
    e. 10 µg/mL stock: Add 1 mL of the 100 µg/mL to 9 mL YPM+0.1% tryptophan broth in a sterile 15 mL conical tube.
    f. 5 µg/mL stock: Add 1 mL of the 50 µg/mL to 9 mL YPM+0.1% tryptophan broth in a sterile 15 mL conical tube.

2. <u>Make Salkowski reagent</u> (2 mL/isolate + 10 mL for standards/blank).

For 40 mL:

1. Add 19.6 mL ddH$_2$O into a clean 200 mL glass bottle.

2. Using a glass pipette, withdraw 19.6 mL of 70% perchloric acid and add to glass bottle with water. WARNING: HIGHLY CORROSIVE. Use appropriate personal protective equipment when handling acid and do this in the chemical hood.

3. Mix 0.8 mL of 0.5 m FeCl$_3$ into the glass bottle. NOTE: If using ferric chloride heptahydrate, use 2.34 g of FeCl$_3$·7H$_2$O in 10 mL of ddH$_2$O to get a 0.5 M solution.

4. Prepare glass test tubes for color metric assay: Label 1.5 mL Eppendorf tubes with the names of the isolates being tested.

5. Preweigh empty tubes and record the weight.

6. Add 1 mL of the 4-day old liquid culture into the labeled tubes.

7. Centrifuge tubes at 20,000 rpm for 10 minutes.

8. Carefully, without disturbing the pellet, extract 950 µL of the supernatant into an appropriately labeled glass test tube. Put the

remaining solution in the 1.5 mL tubes (pellet + remaining supernatant) in a drying oven, cap open, and let completely dry.

9. Once they are dried, weigh the dry weight of the pellet either by weighing the tube and subtracting the original empty weight of the tube or by scraping out the pellet and weighing that (if the tubes were not weighed beforehand).

10. Label an additional 5 test tubes for the IAA standards (100, 50, 20, 10, 5), and pipette 950 µL of the prepared standards into those tubes.

11. Make a blank test tube with 2 mL of plain YPM broth + 0.1% tryptophan.

12. Measure IAA produced by the isolates using the spectrophotometer.

13. Mix the Salkowski reagent well and add 2 mL into each glass test tube. Start a 25-minute timer immediately after adding the Salkowski reagent to the first tube.

14. Incubate at room temperature for 25 minutes. After 25 minutes, the color change in the test tube will indicate the amount of IAA in each tube.

15. Turn on the spectrophotomer and adjust the wavelength to 530 nm.

16. Transfer 1 mL of blank to a cuvette (2 mL if using the 2 mL cuvettes). Use this to blank the spectrophotometer.

17. Transfer 1–2 mL of IAA standards to labeled cuvettes and measure their absorption to create a standard IAA concentration curve later on.

18. Measure absorption for all isolates at OD 530 nm.

Day 6: Data Analysis

1. Generate a standard curve between IAA concentration and absorbance at 530 nm using the data measured from the IAA standard solutions.

2. Use the equation of the line created from this standard curve to calculate IAA concentration based on the absorbance of the isolates being tested.

3. Using the dry weight of the pellet from the isolates, calculate the IAA produced per gram of isolate based on the calculated IAA concentrations.

3.6

SCREENING BACTERIAL ISOLATES FOR SIDEROPHORE PRODUCTION

CONTRIBUTORS: TRI TRAN AND SANA DOTY

This protocol is adapted from the CAS assay for siderophore detection (Louden et al., 2011) and from Schwyn & Neilands (1987).

MATERIALS

LB-CAS agar plates (100 x 100 mM grid square plates)
LB broth
12 mL snap tubes for bacterial liquid cultures
Waste beaker
100% EtOH
Small beaker or jar for EtOH sterilization
Fine-tipped Sharpie

EtOH burner + lighter
P1000 + P200 pipettes + tips
Forceps
Kimwipes
70% EtOH spray
Spectrophotometer
Cuvettes

PROTOCOL

Day 1: Preparation

1. Prepare liquid cultures of bacterial isolates in a fume hood or on the bench-top using a Bunsen burner to maintain a sterile environment.
   a. Label a 12 mL snap tube with isolates to be tested.
   b. Add 2 mL of LB broth to tubes.
   c. Pick a single colony of an isolate using a pipette tip.
   d. Drop the pipette tip into the appropriately labeled tube of LB broth.
   e. Incubates tubes at 28°C on a shaker overnight.
2. Prepare LB-CAS agar plates if not done so (see Chapter 5: Recipes).

Day 2: Experiment

1. Prepare for the experiment.
   a. Clean out and sterilize a biological fume hood with 70% EtOH.
   b. Set up the hood with your waste beaker, EtOH burner, EtOH sterilizing jar/beaker filled with 100% EtOH and forceps, and p1000 tips.
2. Remove LB-CAS plates from the fridge and leave them open in the hood to dry any condensation on plate surfaces and lids.
3. While plates are drying, dilute liquid cultures to make our inoculum.
   a. Remove liquid cultures from the incubator shaker.
   b. Pipette 1 mL of culture into a cuvette.
   c. Pipette 1 mL of plain LB broth into a cuvette. This is our BLANK.
   d. Using the spectrophotometer at a wavelength of 600 nm, blank using our LB blank and then measure the OD of our samples. We are aiming for an OD = 0.2 at 600 nm.
   e. Using the equation $C_1 V_1 = C_2 V_2$, calculate the necessary volume from our initial liquid cultures that we need to create dilutions that will be approximately an OD = 0.2.
   f. Once calculated, take the V_1 amount of the initial liquid culture and add it to LB to create 2 mL of our dilution in new snap tubes or 15 mL conical tubes.
   g. Double-check the dilution concentration on the spectrophotometer using 1 mL of dilution. This will leave us with 1 mL of remaining dilution for our experiment.

1.1	1.2	1.3
2.1	2.2	2.3
3.1	3.2	3.3

FIGURE 3.6.1. Example of plating grid with isolates as technical replicates

4. Once the plates are completely dry, label the plates with a 3 x 3 grid and the date. Three different isolates can be tested per plate, one in each row, with three technical replicates per isolate in each square on that row. Label the rows with the isolate number.

5. Using the wider end of a p1000 tip, punch a hole in the middle of the grid on the plate. Use sterile forceps to carefully remove the circular plug of agar to leave a well in the middle of each square.

6. Pipette 50 μL of the appropriate diluted isolate inoculum in the well of the appropriate grid square. Be sure not to touch the agar to prevent the inoculum from leaking out of the well. Repeat for all three technical replicates.

7. Repeat step 5 for all isolates.

8. Leave plates open in the hood to let dry.

9. Once the plates are dry, incubate at 28°C for 3 days.

Day 5 (3 Days Postinoculation): Scanning Day

1. Remove plates from the incubator.

2. Scan plates using the WinRhizo scanner, connected to the lab computer. Isolates that produce siderophores that chelate iron will have a circular halo around the well.

3. Measure the radius of the halo using ImageJ.

4. Calculate the area of the halo using the formula for the area of a circle (*Area* $=\pi r^2$).

5. Normalize the area by the specific OD of each isolate by dividing the area by the OD.

3.7

SCREENING BACTERIAL ISOLATES FOR PHOSPHORUS SOLUBILIZATION

CONTRIBUTORS: TRI TRAN AND SANA DOTY

This protocol has been adapted from Gupta et al. (1994).

MATERIALS

Pikovskaya plates (100x100 mM grid square plates)	P1000 + P200 pipettes + tips
LB broth	Kimwipes
12 mL snap tubes for bacterial liquid cultures	70% EtOH spray
Waste beaker	Spectrophotometer
Fine-tipped Sharpie	Cuvettes

PROTOCOL

Day 1: Preparation of Liquid Cultures

1. Prepare liquid cultures of bacterial isolates in a biological safety hood or on the benchtop using a Bunsen burner to maintain a sterile environment.
   a. Label 12 mL snap tube with isolates to be tested.
   b. Add 5 mL of LB broth to tubes.

   c. Pick a single colony of isolate using a pipette tip.

   d. Drop the pipette tip into the appropriately labeled tube of LB broth.

   e. Incubates tubes at 28°C on a shaker overnight.

2. Prepare Pikovskaya agar plates if not done so on 100 x 100 mm grid square plates (see Chapter 5).

Day 2: Experiment Day

1. Clean out and sterilize a biological fume hood with 70% EtOH.

2. Remove Pikovskaya plates from the fridge and leave them open in the hood to dry any condensation on plate surfaces and lids.

3. While plates are drying, dilute liquid cultures to make our inoculum.

   a. Remove liquid cultures from incubator shaker.

   b. Pipette 1 mL of culture into a cuvette.

   c. Pipette 1 mL of plain LB broth into a cuvette. This is our BLANK.

   d. Using the spectrophotometer at a wavelength of 600 nm, blank using our LB blank and then measure the OD of our samples. We are aiming for an OD = 0.2 at 600 nm.

   e. Using the equation $C_1 V_1 = C_2 V_2$, calculate the necessary volume from our initial liquid cultures that we need to create dilutions that will be approximately an OD = 0.2.

   f. Once calculated, take the V_1 amount of the initial liquid culture and add it to LB to create 2 mL of our dilution in new snap tubes or 15 mL conical tubes.

   g. Double-check the dilution concentration on the spectrophotometer using 1 mL of dilution. This will leave us with 1 mL of remaining dilution for our experiment.

4. Once the Pikovskaya plates are completely dry, label the plates with a 3 x 3 grid and the date. Three different isolates can be tested per plate, one in each row, with three technical replicates per isolate in each square on that row. Label the rows with the isolate number.

5. Pipette 20 µL of the appropriate diluted isolate inoculum in the middle of the appropriate grid square, making a circular droplet. Repeat for all three technical replicates.

6. Repeat step 5 for all isolates.

7. Leave plates open in the hood to let dry.

8. Once the plates are dry, incubate at 28°C for 5 days.

Day 7 (5 Days Postinoculation): Scanning Day

1. Remove plates from the incubator.
2. Scan plates using the WinRhizo scanner, connected to the lab computer. Isolates that produce metabolites that can solubilize mineral phosphate will have a circular halo around where the droplet was.
3. Measure the radius of the halo using ImageJ.
4. Calculate the area of the halo using the formula for the area of a circle (*Area* $= \pi r^2$).
5. Normalize the area by the specific OD of each isolate by dividing the area by the OD.

3.8

PROTOCOL FOR ASSAY SCREENING ISOLATES WITH IN VITRO DIRECT *RALSTONIA SOLANACEARUM* ANTAGONISM

CONTRIBUTORS: TRI TRAN AND SANA DOTY

MATERIALS

CPG with 0.2% TZC agar on grid square plates
LB broth
14 mL snap tubes for liquid cultures
50 mL Falcon tubes
Ethanol burner + lighter
Glass streakers
Micro pipettes + tips

Ethanol (For sterilizing)
Metal forceps
Waste beaker
Spectrophotometer
Microscope slides (autoclaved)
28°C incubator
Serological pipette + tips

PROTOCOL

Day 1: Streaking Out Ralstonia

1. Streak out your *Ralstonia* strain on CPG plates and incubate at 28°C for 48–72 hours.

   a. Streak out by pipetting 200µL or more on the plates and spreading it evenly using a sterile glass spreader to create a lawn.

   b. <u>Notes</u>:

      i. From glycerol stocks: Takes ~48 hours to grow.

      ii. From water stocks: Takes ~72 hours to grow.

      iii. If *Rs* strain is -GFP tagged, add tet10 to CPG plates.

Day 2: Preparing Liquid Cultures

1. Prepare liquid cultures of isolates to be tested (in a sterile environment).
   a. Fill labeled 14 mL snap tubes with 2 mL of LB broth.
   b. Inoculate tubes with the respective isolates using a colony picked using a sterile pipette tip.
   c. Incubate cultures in a 28°C shaker for 24 hours.
2. Prepare CPG plates for antagonism assay.
   a. Draw a 3 x 3 grid on the bottom surface of the plates using a Sharpie.
   Individual isolates will be tested using one row of 3 squares (3 technical replicates). One plate can be used to test 2 isolates (one on the top row, one on the bottom). The middle row should be left empty to ensure no cross-contamination.

Day 3: Inoculating and Treating Plates

1. Collect *Ralstonia* streaked onto the CPG plate on Day 1.
   a. Add 10 mL of sterile water to the plate.
   b. Carefully scrape the surface of the plate with an autoclaved microscope slide.
   c. Pipette scraped bacterial slurry into a clean 50 mL conical tube using a serological pipette.
2. Prepare ~10 mL of 10^8 CFU/mL *Ralstonia* (OD_{600} ~ 0.2, but this depends somewhat on the spectrophotometer).
   a. Dilute bacterial slurry using sterile water.
   b. Pipette 1 mL of solution into a clean cuvette.
   c. Blank the spectrophotometer (set at 600nm) using sterile water.
   d. Test sample in the spectrophotometer for the OD.
   e. Continue to adjust solution until the appropriate OD has been achieved.
      i. <u>Note</u>: Equation $C_1V_1 = C_2V_2$ can be used to calculate adjustments.

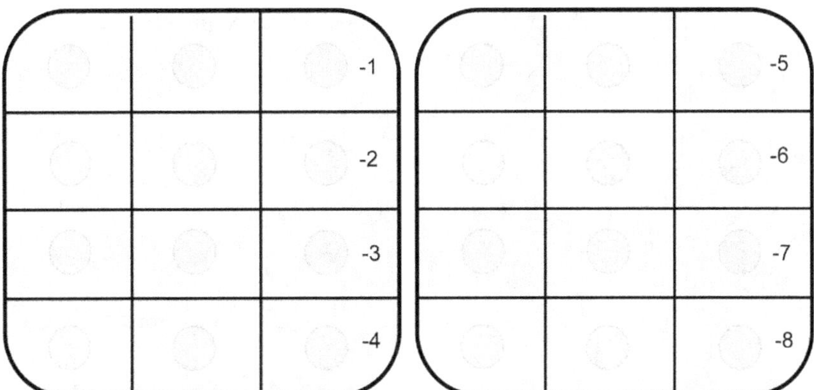

FIGURE 3.8.1. Plating scheme for antagonism assay

3. Prepare tenfold dilutions of the 10^8 CFU/mL *Ralstonia* to confirm accurate concentration on plates.
   a. Add 180µL sterile water to 8 microcentrifuge tubes labeled -1 to -8.
   b. Add 20µL of the 10^8 CFU/mL *Ralstonia* solution to the first tube (labeled -1). Mix by pipetting the solution up and down a few times.
   c. Take 20µL of tube -1 and add to the tube labeled -2 and mix.
   d. Repeat this until all 8 dilutions have been made.
   e. Pipette three 20µL droplets of each dilution on labeled CPG+0.2%TZC plates (Figure 3.8.1).
   f. Let plates air-dry in the hood before incubating at 28°C for 48 hours or until colonies grow.
   g. When counting the colonies, use this formula to calculate the *Ralstonia* inoculum concentration: *Ralstonia* concentration (CFU/ml) = Colony Count on Plate/ (Dilution factor*Volume streaked in mL).

Inoculating Plates for the Assay

1. Prepare and label your mock and *Ralstonia* CPG grid plates.
   a. For the mock solution, use sterile water (preferably autoclaved in the same batch as the water used for the inoculum).
   b. There should be 2 CPG plates total for 2 isolates: one for mock and one for *Ralstonia*, with both isolates for the top and bottom rows.

2. Starting with mock, pipette 200μL of mock solution onto each plate and spread with a sterilized glass spreader.

3. Let plates air-dry in the hood.

4. Repeat steps 5–6 for the *Ralstonia* plates using the 10^8 CFU/mL *Ralstonia* inoculum.

5. Once plates are dry, incubate them at 28°C for 4 hours to allow for establishment.

6. Prepare the isolates for treatment.

   a. Remove the liquid cultures of the isolates to be tested from the 28°C shaker.

   b. Test the OD_{600} of the cultures (*see above: Day 3, Step 2*) and dilute them to an OD_{600} = 0.2.

      i. <u>Note</u>: Blank the spectrophotometer with pure LB broth.

7. After 4 hours, remove the plates from the incubator and return to the hood.

8. Using an inverted sterile p1000 pipette tip, carefully poke holes in the center of the squares along the upper and lower rows of the grids, starting with the mock plates.

   a. If the agar plug gets stuck inside of the tip or plate, use sterilized forceps to remove the plug.

   b. Replace tips between plates.

   c. There should be a total of 9 wells created per plate.

9. Label a mock plate and *Ralstonia* plate with the isolate ID.

   a. Label the top rows of both plates with one isolate ID and the bottom rows with another. The middle row will be left empty.

10. Carefully pipette 50μL of the isolate treatment (OD_{600} = 0.2) into each designated well.

    a. Each of the squares that make up a row will be treated as a technical replicate of the isolate being tested.

    b. The treatment should only go into the well, not on the agar.

    c. Be careful not to touch the plate with the pipette tip to ensure that the treatment does not leak out onto the agar surface.

11. Let the plates dry in the hood for 1 hour before incubating them at 28°C for 72 hours.

    a. Be careful not to disturb the plates to avoid leakage of the treatment from the wells.

Day 6: Scanning the Plates

1. Scan the plates.
   a. The formation of a circular inhibition zone around the wells indicates positive antagonism against *Ralstonia*.
2. Using ImageJ, measure the area of the zone of inhibition.
   a. Scan a ruler to set the scale on ImageJ. This will allow you to measure the zone of inhibition in cm or mM.
   b. Record measurements in an Excel sheet or a lab notebook.

3.9

PCR-BASED SCREENING OF POTENTIAL BIOCONTROL-RELATED GENES IN BACTERIAL ENDOPHYTES

CONTRIBUTORS: TRI TRAN AND SANA DOTY

MATERIALS

Primers used for screening 2,4-Diacetylphloroglucinol (DAPG) biosynthesis potential:

phlD–F: 5'-ACC CAC CGC AGC ATC GTT TAT GAG C-3'

phlD–R: 5'-CCG CCG GTA TGG AAG ATG AAA AAG TC-3'

Primers used for screening hydrogen cyanide (HCN) biosynthesis potential:

hcnAB–F: 5'-TGC GGC ATG GGC GTG TGC CAT TGC TGC CTG G-3'

hcnAB–R: 5'-CCG CTC TTG ATC TGC AAT TGC AGG CC-3'

2x APEX Taq RED Master Mix with 1.5 mM $MgCl_2$

Ice bucket

Nuclease-free water

Isolates to be tested (streaked on a plate)

Thermal cycler

Gel electrophoresis materials

Agarose

1x TAE buffer

RedSafe dye (or some other ethidium bromide equivalent)

Gel box

PROTOCOL

1. Prepare the 20 µL PCR reaction mixture using the protocol below:
   a. 6 µL ddH$_2$O
   b. 1 µL forward primer (1 µM)
   c. 1 µL reverse primer (1 µM)
   d. 10 µL APEX Taq Master Mix
   e. Colony DNA (2 µL made from boiling 1 colony in in 100 µL of nuclease-free water)
2. Place tubes in PCR with 1 min extension time for APEX (may need to adjust for other polymerases) and 30–35 cycles.
3. Run 1% DNA gel to observe bands.

   *Include positive and negative control isolates!

WORKS CITED

Gordon, S. A., & Weber, R. P. (1951). Colorimetric estimation of indoleacetic acid, *Plant Physiology, 26*(1), 192–195.

Gupta, R., Singal, R., Shankar, A, Kuhad, R. C., & Saxena, R. K. (1994). A modified plate assay for screening phosphate solubilizing microorganisms. *Journal of General and Applied Microbiology, 40*(3), 255–260.

Louden, B. C., Haarmann, D., & Lynne, A. M. (2011). Use of blue agar CAS assay for siderophore detection. *Journal of Microbiology and Biology Education, 12*(1), 51–53.

Schwyn, B., & Neilands, J. B. (1987). Universal chemical assay for the detection and determination of siderophores. *Analytical Biochemistry, 160*(1), 47–56.

Tran, T., French, E., & Iyer-Pascuzzi, A. S. (2022). In vitro functional characterization predicts the impact of bacterial root endophytes on plant growth. *Journal of Experimental Botany, 73*(16), 5758–5772.

PART 4

RNA EXTRACTION PROTOCOLS FOR ROOTS

For RNA from large root systems or from many soil-grown roots, we typically use TRIzol™. For RNA from plate-grown tomatoes or seedlings, we use a kit-based method. There are many kits, and in our experience most work well, but we most often use the RNeasy Plant Mini Kit from Qiagen and follow its instructions.

4.1

RNA EXTRACTION FROM ROOT TISSUE WITH TRIzol™

CONTRIBUTOR: DENISE CALDWELL

This protocol was optimized for root tissues from the manufacturer's instructions.

MATERIALS

Ground root tissue (either from plates or soil)
TRIzol™ Reagent (from Invitrogen)
Eppendorf tubes
Chloroform
75% ethanol
RNase-free water
Notes: Label all the tubes you will need for the entire experiment.
Clean bench with 70% ethanol, let dry and then wipe down with Rnase away.

PROTOCOL

The protocol below consists of three procedures: RNA extraction, determining RNA yield, and running a 1% agarose gel.

RNA EXTRACTION

1. Use ~100 mg of root tissue already ground and aliquoted into a 2 mL Eppendorf tube.

2. Add 1 mL of TRIzol™ per 50–100 mg of tissue and vortex for about 10 seconds. Place the tube on the bench (room temp) for 10 minutes and add 200 µL of chloroform. Note: Chloroform is volatile. Add it to the tube in the chemical fume hood.

3. Vortex and incubate at room temp for 2–3 minutes

4. Centrifuge at 12,000 g for 15 minutes at 4°C. Note: Make sure the centrifuge and the rotor are already cold at 4°C; if using a centrifuge in the lab, begin cooling the machine 20 minutes before use. Take clear supernatant (the aqueous phase: RNA should be in this phase. DO NOT DISTURB THE interface or organic phase). Transfer to a new 1.5 mL tube.

5. Add an equal volume of isopropanol (cold) and mix well. Note: It will be ~500—700 µL per 1 mL of TRIzol™. DO NOT VORTEX!

6. Incubate at room temp for 10 minutes.

7. Centrifuge at 12,000 g for 10 minutes at 4°C.

8. Discard the supernatant. Note: Carefully drain it or you will lose the RNA pellet; you can use a pipette.

9. Add 1 ml of 75% EtOH to wash the RNA pellet.

10. Centrifuge 12,000 g for 5 min at 4°C. Discard the supernatant and quick-spin to remove extra solution in the tube. Note: Use the P10 or P200 pipette to take as much of the 75% ETOH out of the tube as possible. Let it dry in the chemical hood for 5–10 min. Note: Do NOT let the pellet overdry; when it is dry and good for resuspending, it should look clear and jelly-like in the tube.

11. Resuspend the pellet in 40 µL of RNase-free water. Note: Warm up the RNase-free liquid at 60°C before use. You may need to incubate in a water bath or heat block set at 55–60°C for 5 minutes (max 10 minutes), as this will help dissolve the RNA better. Store the RNA at -80°C. Note: This is the raw RNA. After this isolation, you need to purify and treat it with DNase.

DETERMINE TOTAL RNA YIELD

Nanodrop (absorbance)
Use 2 µL of the RNA to quantify the RNA concentration and the purity with the 260/280 and 260/230 ratios.
260/280 ~1.8 -2.0
260/230 ~2.0 -2.2

<u>1% AGAROSE GEL</u>

Ideally, run an RNase-free gel. In the absence of that, we have found that cleaning the gel rig and using fresh gel buffers works well to prevent RNA degradation. Use 1–2 μL of the RNA on an agarose gel. You should see 2 bands, clear and not faded.

PART 5

COMMONLY USED AGAR RECIPES FOR TOMATO AND BACTERIAL GROWTH

CPG MEDIA

Media such as Casamino Peptide Agar (CPG) with or without triphenyl tetrazolium chloride (TZC) or modified Sequiera medium South Africa (SMSA) are generally used for growth of pure culture of *R. solanacearum*. (e.g., to retrieve cultures from frozen stocks or for successive plating of cells). These media can also be used for isolation of *R. solanacearum* from fresh symptomatic plants due to the high density of the pathogen in the tissues. We most often use CPG + TZC. For more information see Garcia et al. 2019.

MATERIALS

- Glucose
- Bacto casamino acids
- Bacto yeast extract
- Bacto proteose peptone 3
- Agar, micropropagation Type 1, by Caisson
- ddH$_2$O
- 1% TZC solution (see below for recipe)

PROCEDURE

Making the Media

1. Take a large glass beaker, place it on the stir plate, and add a stir bar. Do not turn on the stir plate.
2. Add the following to the glass beaker at the appropriate concentration:

RECIPE

CHEMICAL	500 ML	800 ML	1000 ML
Glucose (D-glucose, anhydrous, granular)	2.5 g	4.0 g	5.0 g
Bacto proteose peptone 3	5.0 g	8.0 g	10.0 g
Bacto casamino acids	0.5 g	0.8 g	1.0 g
Bacto yeast extract	0.5 g	0.8 g	1.0 g
ddH_2O	450 mL	700 mL	900 mL
Adjust pH to 6.5–7.0 (KOH)			
Agar	8.0 g	12.8 g	16.0 g
Add ddH_2O to desired volume	~40–50 mL	~90–100 mL	~90–100 mL
WHEN COOL to touch, if needed, add 1% TZC	1.0 mL	1.6 mL	2.0 mL

3. Dextrose can be used in the place of glucose if D-glucose is not available (but glucose is preferred).
4. Use a graduated cylinder to add ddH_2O to the glass beaker.
5. Turn on the stir plate and allow media to mix for 5–10 minutes.
6. Weigh out agar and place it into the bottle that will be autoclaved.
7. Use a pH meter to adjust pH (KOH) to between 6.5 and 7.0.
8. Pour the solution into a graduated cylinder and add ddH_2O to the final volume.
9. Add this solution to the final bottle that contains the agar.
10. Adding the stir bar to the final solution, allow to stir for 5 minutes.
11. Autoclave.
12. After autoclaving, allow the media to cool to the touch.
13. Once the media has cooled, add the correct amount of TZC and pour plates in the sterile hood. The addition of TZC allows differentiation of virulent and non-virulent strains.

Large petri dishes are for the inoculum preparation.

Medium petri dishes are for dilution plating.

To Make 1% TZC

1. For 30 mL of solution, measure 0.3 g of TZC.
2. Place the TZC into a 50 mL falcon tube.
3. Add 25 mL of ddH_2O to the falcon tube.
4. Vortex the Falcon tube, then add an additional 5 mL of ddH_2O.
5. Filter sterilize the TZC solution into another 50 mL Falcon tube and wrap the tube below the cap with aluminum foil. Store 1% TZC solution at 4°C.

WATER AGAR (500 ML)

Used for tomato seedling growth on petri plates.

Agar	5 g
ddH_2O	To 500 mL

1. Add agar 900 mL of dH_2O.
2. Mix using a stir bar.
3. Adjust volume to 1,000 mL using a graduated cylinder.
4. Autoclave on standard cycle for 15 minutes.
5. Pour plates in the sterile hood when cool enough to hold the beaker.
6. Store plates in 4°C or at room temperature if using the same day.

1X PBS BUFFER (1 L)

A commonly used buffer to stabilize cells in solution.

NaCl	8 g
KCl	0.2 g
Na_2HPO_4	1.44 g
KH_2PO_4	0.24 g

Add reagents to 900 mL of dH_2O.

1. Mix using a stir bar until everything has dissolved.
2. Adjust volume to 1,000 mL using a graduated cylinder.
3. pH solution to a pH = 7.4 (+/-0.3) with HCl.
4. Either autoclave or filter-sterilize the solution.
5. Store at room temperature.

LB AGAR (1 L)

A nutrient-rich medium commonly used to grow bacteria nonselectively.

Tryptone	10 g
Yeast extract	5 g
NaCl	10 g
Agar	15 g

1. Add reagents (except agar) to 900 mL of dH_2O.
2. Mix using a stir bar until everything has dissolved.
3. Adjust volume to 1,000 mL using a graduated cylinder.
4. pH solution to a pH = 7.0 (+/-0.3).
5. Add media to bottles/flasks and add the agar.
6. Autoclave. Store plates in fridge.

   Notes:

   For LB broth, omit the agar. Broth can be stored at room temperature.

   If adding antibiotics to media, add them to media after autoclaving once media has cooled down to be warm (not hot) to the touch.

PIKOVSKAYA'S AGAR (1 L)

Media used for the detection of phosphate solubilizing bacteria (Paul and Sundra Rao, 1971). You can buy this media or make it with the recipe below.

Yeast extract	0.5 g
Dextrose	10 g
Calcium phosphate	5 g
Ammonium sulfate	0.5 g
Potassium chloride	0.2 g
Magnesium sulfate	0.1 g
Manganese sulfate	0.1 mg
Ferrous sulfate	0.1 mg
Agar	15 g

1. Add reagents to 900 mL of dH_2O.
2. Mix using a stir bar. (NOTE: Phosphate is insoluble and will never dissolve. It will leave media looking milky white and cloudy.)
3. Adjust volume to 1,000 mL using a graduated cylinder.
4. pH solution to pH = 7 (+/-0.3) (w/ KOH).
5. Add media to bottles/flasks and add the agar.

6. Autoclave. Store plates in fridge.

  Notes: Insoluble phosphate will make media a cloud milky color. Before pouring into plates, make sure to mix media to ensure even distribution of phosphate.

YPM BROTH WITH 0.1% TRYPTOPHAN (1 L)

This broth is used to grow liquid cultures of bacteria and yeasts.

Mannitol	25 g
Peptone	3 g
Yeast extract	5 g
Tryptophan	0.1 g

1. Add reagents to 900 mL of dH_2O.
2. Mix using a stir bar until completely dissolved.
3. Adjust volume to 1,000 mL using a graduated cylinder.
4. pH solution to pH = 7.0 (+/-0.3).
5. Autoclave. Store at room temperature.

LB-CAS AGAR PLATES (1 L)

This media has a base of LB with Chromeazurol S (CAS) dye added. Used for the detection of siderophores in bacteria. Originally published by Schwyn and Neilands (1987) and Louden et al. (2011).

Solution 1	0.06 g CAS in 50 mL dH_2O
Solution 2	2.7 mg $FeCl_3 \cdot 6H_2O$ in 10mL 10 mM HCl
Solution 3	0.073 g HDTMA in 40 mL dH_2O

1. Create each solution in glass bottles.

  (Note: Create Solution 2 in a fume hood.)
2. Add 9 mL of Solution 2 to Solution 1. Mix by inverting the bottle.
3. Add mixed solutions to Solution 3. Mix by inverting the bottle. (Note: The final solution should be a dark blue color.) This is the finished CAS dye.
4. Autoclave the CAS dye, along with 900 mL of LB agar.
5. When LB-agar is lukewarm to the touch, add in the CAS dye. Mix media well before pouring. The final media should be a blueish-green color.
6. Solidified plates can be stored in the fridge.

  Note: CAS dye is EXTREMELY SENSITIVE!!! All glassware to be used for both the dye and the LB should be cleaned thoroughly with soap and then rinsed once with dH_2O before *carefully* rinsing all glassware with 6 m HCl to remove any remaining mineral residue. After this rinse, glassware needs to be rinsed thoroughly with dH_2O again before it is ready to be used.

HCl is **HIGHLY CORROSIVE.** This step should only be done wearing full personal protective equipment (lab coat, chemical splash goggles, solvent gloves that are rated for strong acids) and in the presence of the lab safety officer or a trusted researcher. Undergraduates are NOT allowed to clean this glassware. Read the Safety Data Sheet for HCl beforehand and familiarize yourself with proper protocol in case of a spillage. Diluted HCl can be rinsed down the sink.

This media was described by Louden and Lynne (2011) and Schwyn et al. (1987).

KINGS MEDIUM B (1 L)

This media is used for distinguishing fluorescent pseudomonads from nonfluorescent pseudomonads.

Proteose peptone #3	20 g
K_2HPO_4	1.5 g
$MgSO_4 \cdot 7H_2O$	1.5 g
Glycerol	10 mL
Agar	15 g

1. Add proteose peptone #3 to 1 L of dH_2O in a 2 L glass Erlenmeyer flask.
2. Bring the solution to a boil and mix until dissolved.
3. Add all other reagents (except agar) and heat until dissolved.
4. Add agar. Cover the flask with foil and autoclave tape.
5. Autoclave. Plates can be stored in the fridge.

R2A MEDIA (1 L)

This media selects slow-growing bacteria.

Yeast extract	0.5 g
Proteose peptone #3	0.5 g
Casamino acids	0.5 g
Dextrose	0.5 g

Soluble starch	0.5 g
Sodium pyruvate	0.3 g
K_2HPO_4	0.3 g
Magnesium sulfate	0.05 g
Agar	15 g

1. Add reagents to 900 mL of dH_2O.
2. Mix using a stir bar.
3. Adjust volume to 1,000 mL using a graduated cylinder.
4. Pour into a bottle/flask.
5. Add 15 g agar.
6. Autoclave.
7. Wait until lukewarm to pour plates.
8. Solidified plates can be stored in the fridge.

WORKS CITED

García, R. O., Kerns, J. P., & Thiessen, L. (2019). *Ralstonia solanacearum* Species Complex: A quick diagnostic guide. *Plant Health Progress, 20*, 7-13.

Louden, B.C., Haarmann, D., & Lynne, A. M. (2011). Use of blue agar CAS assay for siderophore detection. *Journal of Microbiology and Biology Education, 12*, 51–53.

Paul, N. B., and Sundara Rao, W .V. B. (1971). *Phospate-dissolving Bacteria in the Rhizosphere of Some Cultivated Legumes. Plant and Soil* 35, 127–132.

Schwyn, B., & Neilands, J. B. (1987). Universal chemical assay for the detection and determination of siderophores. *Analytical Biochemistry, 160*(1):47–56.

ABOUT THE AUTHOR

ANJALI S. IYER-PASCUZZI is a professor in the Department of Botany and Plant Pathology at Purdue University. Her research focuses on understanding plant-microbe interactions, particularly the interactions between plants and microbial pathogens at the cellular and molecular level on Earth and during spaceflight. Much of the research in her lab is centered on the tomato root *Ralstonia solanacearum* pathosystem. She has received numerous awards for her research and teaching including the New Innovator Award from the Foundation for Food and Agricultural Research and the Outstanding Graduate Mentor and Teacher Award from Purdue's College of Agriculture. In 2024, she was recognized as one of the twenty-five inspiring women in plant biology by the American Society of Plant Biologists.

ABOUT THE CONTRIBUTORS

DENISE CALDWELL is a PhD student in Anjali Iyer-Pascuzzi's lab. Caldwell earned a BS in horticultural science from Purdue University in 2012 and an MS in botany and plant pathology in 2016. She is an expert in light and electron microscopy and has authored ten publications (as first author or coauthor) in the Iyer-Pascuzzi lab in plant disease research. Caldwell has also been the lead student on a NASA project to examine plant defense responses on the International Space Station.

REBECCA LEUSCHEN-KOHL is a PhD student in Anjali Iyer-Pascuzzi's lab studying pattern-triggered immunity in the roots of Solanaceous species. Leuschen-Kohl graduated with a BS in biochemistry from the University of Nebraska–Lincoln and joined the Iyer-Pascuzzi lab as part of the Quantitative Plant Pathology PhD program. Leuschen-Kohl is an alumna of the Alfred P. Sloan Foundation's Indigenous Graduate Partnership and a recipient of the National Science Foundation Graduate Research Fellowship Program. In 2025, she was awarded the National Science Foundation Postdoctoral Research Fellowship in Biology.

SANA DOTY is a PhD student in Anjali Iyer-Pascuzzi's lab. Mohammad's research focuses on the physiology behind tomato resistance to *Ralstonia solanacearum*. Mohammad received a BS in plant biology from the University of California–Davis in 2020. In 2021, she joined Purdue University as a USDA-NIFA Fellow through the Quantitative Plant Pathology PhD program. While at Purdue, she has mentored numerous undergraduate students.

ABIGAIL ROGERS is a PhD student in Anjali Iyer-Pascuzzi's lab, where she studies effector biology. In 2020 Rogers received a BS in genetics and genomics from the University of Wisconsin–Madison. Currently she is an established early-career author and presenter in the field of molecular plant pathology. She has collaborated with molecular plant pathologists across universities.

TRI TRAN earned a PhD from Purdue University in 2022 after working in Anjali Iyer-Pascuzzi's lab and is currently a postdoc at Michigan State University, where he focuses on microbiomes. While a graduate student, Tran focused on the role of root-associated bacteria in promoting tomato growth.